燕园草木

The Selected Plants of Yan Yuan

许智宏 顾红雅 主编

杨辛题

北京大学出版社
PEKING UNIVERSITY PRESS

图书在版编目（CIP）数据

燕园草木/ 许智宏，顾红雅主编．—北京：北京大学出版社，2011.7
（沙发图书馆·博物志）
ISBN 978-7-301-18854-5

I.燕… II.①许… ②顾… III.①生物学－植物志－北京大学 IV.①Q948.521

中国版本图书馆CIP数据核字（2011）第078990号

书　　　名：	燕园草木
著作责任者：	许智宏　顾红雅　主编
责　任　编　辑：	王立刚
内　文　设　计：	设计·邱特聪 [010-87896477]
标　准　书　号：	ISBN 978-7-301-18854-5/N·0039
出　版　发　行：	北京大学出版社
地　　　址：	北京市海淀区成府路205号　100871
网　　　址：	http://www.pup.cn　电子邮箱：zpup@pup.cn
电　　　话：	邮购部62752015　发行部62750672
	出版部62754962　编辑部62752728
经　销　者：	新华书店
印　刷　者：	北京中科印刷有限公司
开　　　本：	720mm×1020mm　16开本　17.5印张　100千字
版　　　次：	2011年7月第1版　2024年4月第5次印刷
定　　　价：	150.00元

未经许可，不得以任何方式复制或抄袭本书之部分或全部内容。
版权所有，侵权必究
举报电话：010-62752024　电子邮箱：fd@pup.cn

燕园草木
郁郁葱葱
根深叶茂
桃华李芳

辛卯秋日周其凤

序 草木含情

如果从明末米万钟建的"勺园"算起，燕园就已有近四百年历史，期间饱经风雨沧桑，校园内现存的五百多株古树见证了这块园地的变迁。每当出访有植物园的海外大学时，因知道我是学植物学的，都会请我去看看他们的植物园。但我想，燕园就是一座植物园。有山有水，部分园地还保留着半自然生态的状态，加上由于校园绿化和教育的需要，过去几十年，又引进了不少种植物。据资料所载，燕园内约有90多科300多种植物，是北京生物多样性最为丰富的园林之一。正是树木花草，使燕园四季分明。古树峥嵘，百花争艳，又突显出北大的灵气和活力。

北大出版社的王立刚老师（北大哲学系校友）曾在《北大最美的十棵树》中列出了：三角地的柿子林，西门南华表旁的银杏，静园草坪的松树，一院到六院的爬山虎，临湖轩的竹子，未名湖南岸的垂柳，浴室南面的英国梧桐，五四体育馆大门旁的白蜡树，南门主路两旁的槐树，三教足球场东边的白杨树。当然，也许还可列出更多，比如六院的紫藤，中国经济研究中心院内的玉兰，还有"公主楼"前的银杏路，等等。而其中让我们这代人十分留恋的可能要推那片不大的柿子林了，我读书时树长得还不高。如今在百周年纪念讲堂的地方，正是当年容纳了大部分同学用餐的大、小饭厅。每天吃饭，同学们都会经过这片树林。而树林下，也有摆地摊卖书什么的，到夏天西红柿上市，更是在柿树下西红柿堆成山，同学们拿了洗脸盆，五分钱可买一盆，也是一乐！现在是只剩下记忆了。所幸，校园内

还有零星几棵柿树和它的近亲君迁子（黑枣树）。对我而言，另一无法忘怀的是南门内道旁的两排槐树。毕业后，每次回校，我基本上是从南门进的，一见这两排槐树，就觉得像回家了。后来回母校当了校长，每年开学报到那天，又是在这里迎接来自全国各地的新生和他们的家长。一年又一年，这两排槐树不知见证了多少跨入北大大门求学的学子，又目送了多少毕业的学子从这里离开母校的怀抱。

除了古树名木外，那点缀着园林的不同季节的花卉，有栽培的，还有不少野生的，紫色、黄色、白色；当春天来临，看到那路边、坡地上各种小草露出嫩绿的芽尖，还有那绿茵茵的草地，你会感到这里充满活力、激情与希望。赵柏林先生曾书《燕园荷花赞》："出水芙蓉美如玉，阆苑仙葩净无瑕，梗直虚怀洁自好，锦绣燕园有荷花。"赞赏荷花的清雅高洁，很是代表了一批北大读书人的品格。而读厉以宁先生1978年年初所填《木兰花》（校园初春）："湖边残雪风吹去，墙外麦苗青几许，一行燕子报春来，小径花丛闻笑语。黄昏忽又潇潇雨，乍暖还寒何足虑，隆冬已尽再难回，历史无情终有序。"又见即使在风雨缥缈之中，人有情，草木也有情。北大人的风骨、精神，可以在山水草木之间找到身影。每个北大人，当你离开燕园，哪怕是暂时的，也会梦魂依旧，留恋、思念。季老在《汉城忆燕园》一文中所表露出的是在异国对燕园山水花木的怀念，"虽已深秋，塘中荷叶，依然浓绿，秋风乍起，与水中的倒影共同摇摆。塘畔垂柳，依然烟笼一里堤。小山上的黄栌尚未变红，而丰华月季，却真名副其实，红艳怒放，胜于二月春花"。如此这般割不断的燕园情，诚如谢冕老师在《永远的校园》中所说，每一位北大学子就像"一颗蒲公英小小的种子"，"选择了燕园一片土"，从此在这里发芽、成长。燕园接纳了北大，使北大在这里延伸、传承。

从校长岗位上退下来已两年多了。当时就有生命科学研究院的老师问过我，退下不当校长后想做些什么事。我说我想用两年时间好好看看校园里的一草一木，用相机照下来，编一本介绍燕园植物的书，让我们在燕园工作学习的老师同学们更好地了解我们校园的植物，认识园中的花草树木，同时也可向访问北大的外宾更好地介绍我们的校园。当然，这本书更大的寄托是：希望我们北大学子更好地珍爱我们的家园，保护好燕园的一草一木。欣慰的是这一想法很快得到校内不少老师、同学、摄影爱好者的支持。当年教我植物分类学的汪劲武教授提供了不少资料和具体的指导；生科院顾红雅教授做了大量组织协调工作，并提供了所有物种的英文描述；中国科学院植物研究所的李振宇研究员对收进本书的每个物种进行了校对；哲学系刘华杰教授不仅提供了精美的校园植物照片，还参与了小短文的撰写和最后的定稿工作；国际合作部张莹女士、生科院的魏丽萍、饶广远、瞿礼嘉等教授提供了大量的校园植物照片，中科院植物所的刘夙同学（北大校友）不但提供了他在校期间拍摄的所有的植物照片，还参与了本书植物的中文描述工作，并为不少物种撰写了小短文；生科院教学中心的孟世勇老师提供了所有物种的中文描述；参与本书编著的几位北大同学，如吴岚、赵瑞白、张慧婷、江都、蔡乐，或提供了校园植物照片，或收集或撰写了与燕园植物相关的文章；生科院顾孝诚教授利用春节的休息时间对本书的英文描述进行了修改，并提出了很好的建议；北大出版社的王立刚老师参与了这本书的协调工作，并在编辑方面做了大量的工作；李岩松副校长和国际合作部的夏红卫部长对本书的出版提供了具体的意见和支持；外籍教师 Dr. Richard Wood 和 Mrs. Ruth Ellen Wood 也参与了英文稿的修改工作。正是在大家的通力合作和支持下，才使此书有望今年校庆期间出版，在此一并致谢。我也要特别感谢杨辛教授，当我登门拜访他，说明出版此书的用意时，他欣然同意为本书题名。需要说明的是，因是出彩图版，这次我们并没有把燕园所有的植物都收进来。校园几经变迁，有些植物还没有合适的照片，加上时间紧迫，只得留待以后修订。我也很高兴地知道生科院吕植教授正在编写一本此书的姐妹篇——《燕园动物》，期盼该书也能尽早出版。

2011年2月28日

序号	中文名	学名	页码	序号	中文名	学名	页码
001	迎春花	*Jasminum nudiflorum*, Oleaceae	1	047	抱茎小苦荬	*Ixeridium sonchifolium*, Asteraceae	73
002	山桃	*Amygdalus davidiana*, Rosaceae	2	048	中华小苦荬	*Ixeridium chinense*, Asteraceae	74
003	连翘	*Forsythia suspensa*, Oleaceae	4	049	地黄	*Rehmannia glutinosa*, Scrophulariaceae	76
004	蜡梅	*Chimonanthus praecox*, Calycanthaceae	6	050	牡丹	*Paeonia suffruticosa*, Paeoniaceae	78
005	榆树	*Ulmus pumila*, Ulmaceae	8	051	芍药	*Paeonia lactiflora*, Paeoniaceae	80
006	垂柳	*Salix babylonica*, Salicaceae	10	052	巴天酸模	*Rumex patientia*, Polygonaceae	82
007	旱柳	*Salix matsudana*, Salicaceae	12	053	日本小檗	*Berberis thunbergii*, Berberidaceae	83
008	毛白杨	*Populus tomentosa*, Salicaceae	14	054	楸	*Catalpa bungei*, Bignoniaceae	84
009	早开堇菜	*Viola prionantha*, Violaceae	16	055	棣棠花	*Kerria japonica*, Rosaceae	86
010	紫花地丁	*Viola philippica*, Violaceae	17	056	黄刺玫	*Rosa xanthina*, Rosaceae	87
011	三色堇	*Viola tricolor*, Violaceae	18	057	洋槐	*Robinia pseudoacacia*, Fabaceae	88
012	杜仲	*Eucommia ulmoides*, Eucommiaceae	19	058	紫荆	*Cercis chinensis*, Fabaceae	90
013	诸葛菜	*Orychophragmus violaceus*, Brassicaceae	20	059	大花野豌豆	*Vicia bungei*, Fabaceae	92
014	东京樱花	*Cerasus yedoensis*, Rosaceae	22	060	紫藤	*Wisteria sinensis*, Fabaceae	93
015	日本晚樱	*Cerasus serrulata* var. *lannesiana*, Rosaceae	24	061	虞美人	*Papaver rhoeas*, Papaveraceae	96
016	郁李	*Cerasus japonica*, Rosaceae	25	062	鸢尾	*Iris tectorum*, Iridaceae	98
017	桃	*Amygdalus persica*, Rosaceae	26	063	马蔺	*Iris lactea*, Iridaceae	100
018	榆叶梅	*Amygdalus triloba*, Rosaceae	28	064	黄菖蒲	*Iris pseudacorus*, Iridaceae	101
019	杏	*Armeniaca vulgaris*, Rosaceae	30	065	德国鸢尾	*Iris germanica*, Iridaceae	102
020	玉兰	*Yulania denudata*, Magnoliaceae	32	066	独行菜	*Lepidium apetalum*, Brassicaceae	103
021	紫玉兰	*Yulania liliiflora*, Magnoliaceae	34	067	早园竹	*Phyllostachys propinqua*, Poaceae	104
022	拟南芥	*Arabidopsis thaliana*, Brassicaceae	35	068	乳浆大戟	*Euphorbia esula*, Euphorbiaceae	106
023	紫丁香	*Syringa oblata*, Oleaceae	36	069	乌头叶蛇葡萄	*Ampelopsis aconitifolia*, Vitaceae	107
024	皱皮木瓜	*Chaenomeles speciosa*, Rosaceae	38	070	异穗薹草	*Carex heterostachya*, Cyperaceae	108
025	蒲公英	*Taraxacum mongolicum*, Asteraceae	40	071	臭草	*Melica scabrosa*, Poaceae	109
026	元宝枫	*Acer truncatum*, Aceraceae	42	072	夏至草	*Lagopsis supina*, Lamiaceae	110
027	桑	*Morus alba*, Moraceae	44	073	蛇莓	*Duchesnea indica*, Rosaceae	111
028	构树	*Broussonetia papyrifera*, Moraceae	46	074	刺儿菜	*Cirsium setosum*, Asteraceae	112
029	香茶藨子	*Ribes odoratum*, Saxifragaceae	48	075	平车前	*Plantago depressa*, Plantaginaceae	113
030	点地梅	*Androsace umbellata*, Primulaceae	49	076	圆叶鼠李	*Rhamnus globosa*, Rhamnaceae	114
031	黄栌	*Cotinus coggygria*, Anacardiaceae	50	077	灰栒子	*Cotoneaster acutifolius*, Rosaceae	115
032	黄杨	*Buxus sinica*, Buxaceae	52	078	二球悬铃木	*Platanus acerifolia*, Platanaceae	116
033	华山松	*Pinus armandii*, Pinaceae	53	079	毛梾	*Cornus walteri*, Cornaceae	118
034	白皮松	*Pinus bungeana*, Pinaceae	54	080	红瑞木	*Cornus alba*, Cornaceae	119
035	银杏	*Ginkgo biloba*, Ginkgoaceae	56	081	毛洋槐	*Robinia hispida*, Fabaceae	120
036	水杉	*Metasequoia glyptostroboides*, Taxodiaceae	58	082	火棘	*Pyracantha fortuneana*, Rosaceae	121
037	白杄	*Picea meyeri*, Pinaceae	60	083	欧洲荚蒾	*Viburnum opulus* subsp. *opulus*, Adoxaceae	122
038	油松	*Pinus tabuliformis*, Pinaceae	62	084	互叶醉鱼草	*Buddleja alternifolia*, Loganiaceae	123
039	圆柏	*Juniperus chinensis*, Cupressaceae	64	085	杠柳	*Periploca sepium*, Asclepiadaceae	124
040	侧柏	*Platycladus orientalis*, Cupressaceae	65	086	白杜	*Euonymus maackii*, Celastraceae	125
041	胡桃	*Juglans regia*, Juglandaceae	66	087	红丁香	*Syringa villosa*, Oleaceae	126
042	七叶树	*Aesculus chinensis*, Hippocastanaceae	68	088	雪柳	*Fontanesia phillyraeoides* subsp. *fortunei*, Oleaceae	127
043	美国红梣	*Fraxinus pennsylvanica*, Oleaceae	69	089	红花锦鸡儿	*Caragana rosea*, Fabaceae	128
044	栓皮栎	*Quercus variabilis*, Fagaceae	70	090	锦带花	*Weigela florida*, Caprifoliaceae	129
045	斑种草	*Bothriospermum chinense*, Boraginaceae	71	091	金银忍冬	*Lonicera maackii*, Caprifoliaceae	130
046	附地菜	*Trigonotis peduncularis*, Boraginaceae	72	092	君迁子	*Diospyros lotus*, Ebenaceae	131

093	柿	*Diospyros kaki*, Ebenaceae	132		140	臭 椿	*Ailanthus altissima*, Simaroubaceae	200
094	毛泡桐	*Paulownia tomentosa*, Scrophulariaceae	134		141	香 椿	*Toona sinensis*, Meliaceae	202
095	三裂绣线菊	*Spiraea trilobata*, Rosaceae	135		142	梓	*Catalpa ovata*, Bignoniaceae	204
096	山 楂	*Crataegus pinnatifida*, Rosaceae	136		143	黄金树	*Catalpa speciosa*, Bignoniaceae	206
097	流苏树	*Chionanthus retusus*, Oleaceae	138		144	忍 冬	*Lonicera japonica*, Caprifoliaceae	208
098	大花糯米条	*Abelia×grandiflora*, Linnaeaceae	140		145	蜀 葵	*Alcea rosea*, Malvaceae	210
099	山皂荚	*Gleditsia japonica*, Fabaceae	141		146	梧 桐	*Firmiana simplex*, Sterculiaceae	211
100	蝟 实	*Kolkwitzia amabilis*, Caprifoliaceae	142		147	紫 薇	*Lagerstroemia indica*, Lythraceae	212
101	杂种鹅掌楸	*Liriodendron chinense×L. tulipifera*, Magnoliaceae	144		148	狗尾草	*Setaria viridis*, Poaceae	214
102	卫 矛	*Euonymus alatus*, Celastraceae	146		149	旋覆花	*Inula japonica*, Asteraceae	216
103	太平花	*Philadelphus pekinensis*, Saxifragaceae	147		150	泽 芹	*Sium suave*, Apiaceae	217
104	荇 菜	*Nymphoides peltata*, Menyanthaceae	148		151	一串红	*Salvia splendens*, Lamiaceae	218
105	小花扁担杆	*Grewia biloba* var. *parviflora*, Tiliaceae	150		152	马 蓼	*Polygonum lapathifolium*, Polygonaceae	219
106	月季花	*Rosa chinensis*, Rosaceae	151		153	红 蓼	*Polygonum orientale*, Polygonaceae	220
107	玫 瑰	*Rosa rugosa*, Rosaceae	152		154	乌蔹莓	*Cayratia japonica*, Vitaceae	221
108	女 贞	*Ligustrum lucidum*, Oleaceae	154		155	槐	*Sophora japonica*, Fabaceae	222
109	暴马丁香	*Syringa reticulata*, Oleaceae	155		156	鸭跖草	*Commelina communis*, Commelinaceae	224
110	萱 草	*Hemerocallis fulva*, Liliaceae	156		157	茜 草	*Rubia cordifolia*, Rubiaceae	225
111	酢浆草	*Oxalis corniculata*, Oxalidaceae	158		158	求米草	*Oplismenus undulatifolius*, Poaceae	228
112	草地早熟禾	*Poa pratensis*, Poaceae	159		159	野大豆	*Glycine soja*, Fabaceae	229
113	风花菜	*Rorippa globosa*, Brassicaceae	160		160	短尾铁线莲	*Clematis brevicaudata*, Ranunculaceae	230
114	泥胡菜	*Hemistepta lyrata*, Asteraceae	161		161	葎 草	*Humulus scandens*, Cannabaceae	231
115	半 夏	*Pinellia ternata*, Araceae	162		162	益母草	*Leonurus japonicus*, Lamiaceae	232
116	萹 蓄	*Polygonum aviculare*, Polygonaceae	164		163	玉 簪	*Hosta plantaginea*, Liliaceae	234
117	茴茴蒜	*Ranunculus chinensis*, Ranunculaceae	165		164	萝 藦	*Metaplexis japonica*, Asclepiadaceae	236
118	马齿苋	*Portulaca oleracea*, Portulacaceae	166		165	紫茉莉	*Mirabilis jalapa*, Nyctaginaceae	238
119	藜	*Chenopodium album*, Chenopodiaceae	167		166	虎尾草	*Chloris virgata*, Poaceae	239
120	铁苋菜	*Acalypha australis*, Euphorbiaceae	168		167	芦 苇	*Phragmites australis*, Poaceae	240
121	粉花绣线菊	*Spiraea japonica*, Rosaceae	169		168	盒子草	*Actinostemma tenerum*, Cucurbitaceae	242
122	枣	*Ziziphus jujuba*, Rhamnaceae	170		169	多花胡枝子	*Lespedeza floribunda*, Fabaceae	244
123	石 榴	*Punica granatum*, Punicaceae	172		170	白果毛核木	*Symphoricarpos albus*, Caprifoliaceae	245
124	一叶萩	*Flueggea suffruticosa*, Euphorbiaceae	174		171	龙 葵	*Solanum nigrum*, Solanaceae	246
125	蒙 椴	*Tilia mongolica*, Tiliaceae	175		172	凤尾兰	*Yucca gloriosa*, Agavaceae	247
126	牵 牛	*Ipomoea nil*, Convolvulaceae	176		173	野艾蒿	*Artemisia lavandulifolia*, Asteraceae	248
127	打碗花	*Calystegia hederacea*, Convolvulaceae	177		174	牛 膝	*Achyranthes bidentata*, Amaranthaceae	249
128	圆叶牵牛	*Ipomoea purpurea*, Convolvulaceae	180		175	黄花蒿	*Artemisia annua*, Asteraceae	250
129	田旋花	*Convolvulus arvensis*, Convolvulaceae	181		176	翠 菊	*Callistephus chinensis*, Asteraceae	252
130	栾 树	*Koelreuteria paniculata*, Sapindaceae	182		177	牛膝菊	*Galinsoga parviflora*, Asteraceae	253
131	荆 条	*Vitex negundo*, var. *heterophylla*, Verbenaceae	184		178	鳢 肠	*Eclipta prostrata*, Asteraceae	254
132	枸 杞	*Lycium chinense*, Solanaceae	186		179	婆婆针	*Bidens bipinnata*, Asteraceae	255
133	厚萼凌霄	*Campsis radicans*, Bignoniaceae	187		180	狼杷草	*Bidens tripartita*, Asteraceae	256
134	栝 楼	*Trichosanthes kirilowii*, Cucurbitaceae	188		181	海州常山	*Clerodendrum trichotomum*, Verbenaceae	257
135	华北珍珠梅	*Sorbaria kirilowii*, Rosaceae	190		182	羽叶栾树	*Koelreuteria bipinnata*, Sapindaceae	258
136	莲	*Nelumbo nucifera*, Nelumbonaceae	191		183	山马兰	*Kalimeris lautureana*, Asteraceae	259
137	东方香蒲	*Typha orientalis*, Typhaceae	194		184	甘 菊	*Dendranthema lavandulifolium*, Asteraceae	260
138	木 槿	*Hibiscus syriacus*, Malvaceae	196		185	雪 松	*Cedrus deodara*, Pinaceae	261
139	爬山虎	*Parthenocissus tricuspidata*, Vitaceae	198					

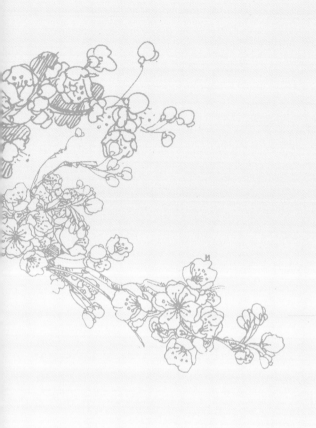

迎春花

木犀科素馨属
Jasminum nudiflorum
(winter jasmine)
Oleaceae

落叶灌木，原产于我国北部及中部。枝条细长，呈拱形下垂生长，侧枝健壮，四棱形，绿色。三出复叶对生，小叶卵状椭圆形，表面光滑，全缘。花期3—5月，可持续50天之久；花单生于叶腋间，花冠高脚杯状，鲜黄色，顶端通常6裂，或成复瓣。为早春开花的中国名贵花卉之一，不仅花色端庄秀丽，而且具有不畏寒威、不择风土、适应性强等特点，历来为人们所喜爱。勺园北部的土山南坡和鸣鹤园湖西岸有大片栽培，春天花开时甚为壮观；未名湖北岸及其他地点也有零星种植。

Deciduous shrub, native to central China. This species blooms very early in spring. This plant is famous for tolerating cold temperatures and for its brightly colored flowers in early spring. It is widely cultivated in China, and the whole plant is used medicinally. Two large populations are found on campus – one north of Shao Yuan, and one west of Ming He Yuan.

迎春花并不去抢开春第一朵的位子，只是默默蓄着力，在最初的惊喜几近淡却之时，倏然怒放。从佝偻枯枝上绽放的明丽的黄，最热烈偏又最单纯，忘我般殷切地吹响春的号角，仿佛坚信她那点微末的光明能够点燃整个春季的激情——哪怕之后凋零在满园花开的华年锦时，也要执守那个传承千年的春祭的仪式。（蔡乐）

山桃

蔷薇科桃属
Amygdalus davidiana (David peach)
Rosaceae

落叶小乔木，分布于我国南北各省。树皮暗紫色，光滑而有光泽。单叶互生，在芽内为对折状，卵圆状披针形，边缘有细锐锯齿，两面无毛。花期3-4月，早于桃和碧桃；花两性，2朵并生，先叶开放，无柄，萼片平展，花瓣5，白色或浅红色；雄蕊多数，花药紫色。果期7月，核果，球形，果核小，有凹沟。未名湖边栽培多，与垂柳共同构成早春湖畔"桃红柳绿"景观。

Small deciduous tree, native to central, North and Northeast China. The seeds of this species are famous traditional medicine for treating some eye diseases. It is planted alternately with willow trees around Wei Ming Lake, and the blooming trees are one of the attractions in early spring on campus.

柳色正新，有柳处不能无山桃。山桃开了，便洗去最后的冬恹，所有过去的冰霜雨雪在晶莹的花瓣的光彩中，再严酷也好似一场宿醉而已，春风一如那岁月静好的承诺。（蔡乐）

连翘

木犀科连翘属
Forsythia suspensa
(weeping forsythia)
Oleaceae

又称黄寿丹。落叶灌木，原产于我国中部和北部。枝直立或下垂，稍开展。叶对生，单叶或羽状三出复叶，叶片边缘有锐锯齿。花期3–4月，花两性，先叶开放，花萼裂片4，长于花冠管，花冠黄色，裂片4；雄蕊2。果期5–6月，蒴果，椭圆形，2瓣裂。为著名观赏植物和药用植物，校园各处栽培甚多，未名湖西北岸的几株长势甚好，花季时非常灿烂。

Deciduous shrub, endemic to central and North China. This species is widely cultivated for its bright yellow flowers in early spring. It is also used medicinally as an antipyretic and antidote. Several big shrubs on the northwest side of Wei Ming Lake are one of the attractions in spring.

连翘和迎春花早春绽放。连翘果实常常和忍冬搭配成药，如北大学生再熟悉不过的"双黄连口服液"（"连"即连翘，"双"则是忍冬别名"双花"）和"维C银翘片"（"银"指忍冬另一别名"金银花"）。连翘在每年深秋众芳摇落时都有反季开花现象，有的植株上的个别花芽会提前好几个月绽放，颇有北大人的一种特立独行的气质。（刘夙）

蜡梅

蜡梅科蜡梅属
Chimonanthus praecox
(Japanese allspice)
Calycanthaceae

落叶灌木，原产于我国中部和东部，现广为栽培。枝灰色，具疣状皮孔。单叶对生，全缘，具短柄；无托叶。花期在北京为2–3月，花两性，单生，先叶开放，花被片多数，无花萼与花瓣之分，均为花瓣状，蜡黄色；雄蕊多数。果期9–10月。著名观赏植物，品种较多，未名湖北岸有栽培。

Deciduous shrub or small tree, native to the central and East China. This species has a long history of cultivation in China for its sweetly fragrant yellow flowers which bloom in late winter. The leaves, roots, flowers, and seeds are used medicinally. Several trees are planted north of Wei Ming Lake, and they only flower when the winter is not too cold.

燕园草木

蜡梅是家喻户晓的名花，因其花为黄色，像蜂蜡的颜色，所以叫"蜡梅"，而非腊梅。蜡梅形如一朵朵小小的垂灯。其香幽深清远，隔墙绕树香远益芬。（刘凤）

榆树

榆科榆属
Ulmus pumila (Chinese elm)
Ulmaceae

又称家榆。落叶乔木,分布于我国长江以北各省及西藏。树皮粗糙,纵裂。单叶互生,卵形或椭圆状披针形,基部稍偏斜,叶缘有重锯齿或单锯齿。花期3月,花先叶开放,两性,多数簇生成聚伞花序,花药紫色。果期4-5月,翅果通称榆钱,倒卵形,先端凹陷,中央为种子,周边具一圈翅。校园极多,临湖轩西南草地、俄文楼西边、燕南园内均有古树。

Deciduous tree, native to most part of China, Korea, Mongolia, East Russia, and central Asia. This species is cultivated throughout China, and in other countries in Asia and North America. It is a rapidly growing tree, and often cultivated in gardens and parks. It is a common tree species on campus, and several big trees over 100 years old are found near the statue of Professor Li Dazhao, and in Yan Nan Yuan.

榆树的叶子和幼果（榆钱）都可以食用。嵇康说豆子吃多了让人身体沉重，榆钱吃多了让人昏睡不醒，从此留下了一个颇为僻雅的成语"榆瞑豆重"，形容人本性难改。（刘凤）

垂柳

杨柳科柳属
Salix babylonica (weeping willow)
Salicaceae

落叶乔木，我国广泛分布。雌雄异株。小枝褐色，细长，下垂。单叶互生，条状披针形或狭披针形，边缘有细锯齿，两面无毛。花期3-4月，雄花序与雌花序均为柔荑花序，与叶同放；雌花只有1枚腺体（可与绦柳区别）。果期4月，果实为蒴果；种子具丝状毛，为北京春天飞絮的来源之一。为优美的绿化和园林树种，未名湖边栽培多，春季与山桃、碧桃形成"桃红柳绿"景观。

Deciduous tree, native to North China. This species is cultivated throughout China, and traded along the silk road to Northwest Asia and Europe. It is famous for its strongly pendulous branches and twigs, and has long been used as an ornamental tree worldwide. It is also grown for wood production, weaving wicker baskets, and shelterbelts. The weeping willow trees are mostly cultivated along the Wei Ming Lake, making a beautiful attraction in spring and fall seasons.

北大里最早的勺园在清代有所谓"风烟里"的旧称。勺园湮灭已久,未名湖畔的株株垂柳仲春初夏之时,绿如烟霭,颇助追古之想。(王立刚)

旱柳

杨柳科柳属
Salix matsudana (Chinese willow)
Salicaceae

落叶乔木，原产于我国，北方栽培多。雌雄异株。小枝黄色，光滑，不下垂或下垂（后者名为绦柳）。单叶互生，披针形，边缘有明显细锯齿，下面灰白色。花期4月，雄花序和雌花序均为柔荑花序，与叶同放；雌花具2枚腺体。果期5月，果实为蒴果；种子具丝状毛，成熟后随风飘散，为北京春天飞絮的来源之一。绿化树种，西校门内有多株老树。

Deciduous tree, native to China. The Chinese willow has been introduced into many areas as an ornamental tree, including Australia, Europe and North America. Several old trees can be seen on the western part of campus.

比起垂柳来，旱柳过于刚硬，失之婀娜，也因此失去了骚人墨客的青睐。46楼下就有一棵旱柳，秋冬时残存的柳枝钢丝般在风中一团乱舞，在男生楼灰白色外墙的映衬下显得分外凄厉。然而就在春风初至时，它却好似一夜间添了新绿，宛如小姑娘的发辫，俏皮而飞扬。我觉得旱柳有一种北方大汉般的性格，粗犷豪放的外表下，却有一颗敏感柔和的心。若非如此，又怎能经受北国这短暂而干旱的春？（蔡乐）

毛白杨

杨柳科杨属
Populus tomentosa (Chinese white poplar)
Salicaceae

落叶乔木，雌雄异株，为我国特有物种。幼树皮灰白色；小枝幼时被白毛，后渐脱落。单叶互生，三角状卵形，边缘具波状齿，下面常被绒毛。花期3月下旬至4月上旬，雄花序和雌花序都为多毛的下垂柔荑花序，先于叶开放。果期4月下旬至5月上旬，果实为蒴果；种子上有白色棉毛，成熟后随风飘散，为北京春天飞絮的主要来源。耐寒、耐旱，生长迅速，是北京和华北的重要木材树种，也是常见的行道树和公园绿化以及防风固沙林树种。第二体育馆附近多有栽培，理发店北侧亦有一株老树。

Deciduous tree, endemic to China. This species is used for wood pulp, timber, and also commonly planted along streets, and as an ornamental. The female trees produce a large amount of hairy seeds which could cause allergies. The plants are found around the Gymnasium No.2 on campus.

毛白杨真像植物中的仁者。姿态伟岸、正直，外表朴质。又如仁者般的外冷内热，早春杨絮飘逸时，应该就是孔夫子当年春服既成沂水沐浴的时候；到了秋冬，那一地厚厚的疏松的落叶，又给校园小动物提供了多少庇护。（吴岚）

早开堇菜

菫菜科菫菜属
Viola prionantha (serrate violet)
Violaceae

多年生草本，分布于我国华北和东北各省。无地上茎。单叶基生，长圆状卵形或卵形，托叶基部和叶柄合生，叶柄上部具翅。花果期从春季到秋季（始花较紫花地丁早），但夏季为闭锁花，仅春秋两季为正常花；花瓣紫色，偶尔有白斑，具较粗短的距。果实为蒴果，椭圆形，3瓣裂。校园内野生极多，凡路边草地林下皆可见，常成片生长。早开堇菜与紫花地丁最显著的区别在于早开堇菜叶子较宽，而紫花地丁叶子细长。

Perennial herb, native to North and Northeast China, Korea, and Russia. This species blooms very early in spring. The whole plant is used medicinally. It is commonly found on campus, usually in large populations.

紫花地丁

堇菜科堇菜属
Viola philippica (Philippine violet)
Violaceae

多年生草本，分布于我国北方。无地上茎，地下茎很短，根白色。单叶基生，狭披针形，开花后增大后形状更明显，边缘具圆齿，叶柄具狭翅；托叶钻状三角形，有睫毛。花期4–5月，花瓣紫堇色，具细管状距，直或稍上弯。果期4–8月，蒴果，长圆形，3瓣裂。校园各草地、路边、林下习见，为春天早开花的草本之一。

Perennial herb, native to China, and other countries in East and Southeast Asia. This is a common species blooming early in spring. The whole plant is used medicinally, and it is a good species for ground cover. It is very commonly seen on campus.

三色堇

董菜科董菜属
Viola tricolor (heartsease or pansy)
Violaceae

草本，原产于欧洲，有很多栽培品种，我国各地广泛栽培。单叶，互生，基生叶具长柄，茎生叶长圆状卵形，边缘具稀疏的圆齿；托叶羽状深裂。花期4–7月，花大，单生叶腋，通常每花有紫、白、黄三色；萼片5，绿色，基部附属物具不整齐边缘；花瓣5，上瓣深紫色，侧瓣及下瓣均为三色，有紫色条纹。果期5–8月，蒴果，椭圆形，无毛。校园花坛常有栽培。

Annual biennial, or perennial herb, native to Europe. This species has been hybridized and bred into many cultivars which are the frequently used plants for gardening. It has been used as a herbal medicine to treat epilepsy, asthma, skin diseases, and chest complaints. Several cultivars are commonly planted on campus.

杜仲

杜仲科杜仲属
Eucommia ulmoides (eucommia)
Eucommiaceae

落叶乔木，原产于我国长江流域，全国各地多有栽培。树皮灰色，小枝光滑，具片状髓。皮、枝、叶和果翅均含胶质。单叶互生，卵状椭圆形，基部宽楔形，边缘有锯齿；叶柄长1-2厘米。花期4-5月，花单性，雌雄异株，无花被；果期9月，具翅小坚果，扁平，先端下凹。其树皮为名贵中药，是为数不多的树皮被环形剥离后能再生的树种，为国家二级保护植物。老生物楼前和四院南侧有数株栽培。

Deciduous tree, endemic to central China. This species is rare in the wild, but widely cultivated. Its bark, which contains aucubin, is used medicinally as an invigorator, a tonic for arthritis, and for reducing blood pressure. It is one of the 50 fundamental herbal medicines in Chinese herbology. Its solidified latex is also used for lining pipes, insulating electric cables, and for filling teeth. Several big trees are found in front of the former Biology Building.

杜仲没有艳丽的花果，将它与其他树种混种，你是看不出它有何出众之处的。然而《神农本草经》说它是"多服久服不伤人"的上品。杜仲还非常顽强，是不怕剥皮的，甚至不怕"环状剥皮"，草木生命力之强，杜仲可说是明证。（顾红雅）

诸葛菜

十字花科诸葛菜属
Orychophragmus violaceus
(violet orychophragmus)
Brassicaceae

又称二月蓝（或讹为"兰"）。一至二年生草本，我国长江以北分布广泛。全株无毛，茎单一，直立。单叶基生或在茎上互生，叶形变化大，下部叶大头羽状分裂，上部叶基部耳状，抱茎。花果期4-6月，花两性，紫色或褪为白色，花瓣开展，4枚，十字形；雄蕊6，4长2短（即所谓"四强雄蕊"）。长角果线形，具4棱。为北京早春开花植物之一，校园极多见。

Annual or perennial herb. This species is native in central, North, Northeast and East China, and Korea; naturalized in Japan. It is an early-spring flowering plant, and can adapt to various habitats. The rosette and young inflorescence are edible. It is probably the most common natural plant on campus.

那是野生的花，浅紫掺着乳白，仿佛有一层亮光从花中漾出，随着轻拂的微风起伏跳动，充满了新鲜，充满了活力，充满了生机。简直让人不忍走开。（宗璞）

东京樱花

蔷薇科樱属
Cerasus yedoensis
(Yoshino cherry)
Rosaceae

落叶乔木，原产于日本，我国各地有栽培。叶片椭圆状卵形，边缘有尖锐的单或重锯齿，多少带刺芒状。花期3月底至4月初，伞房状总状花序，花两性，先叶开放，萼筒管状，外有短柔毛，萼片边缘有细齿，花瓣5，顶端内凹，初时淡红色，后转白色。著名观赏植物。

Deciduous tree. This species occurs as a natural hybrid in Japan, and is one of the most popular and widely cultivated flowering cherries in northern temperate climates. The blossoms of Yoshino cherry near the statue of Cervantes on west campus, create one of the most attractive scenes in early spring. These trees were planted in 2005 as the gift from Waseda University.

盛开于春日初沓之时,花期短暂,怒放之后瞬即凋零,唯美而刚烈,暗合着日本文化的精神,故为日本国花,塞万提斯塑像东侧有几株"早大樱",是日本早稻田大学2005年赠送给我校的礼物。(李洪权)

日本晚樱

蔷薇科樱属
Cerasus serrulata var. *lannesiana*
(Japanese late cherry)
Rosaceae

落叶乔木，我国广为栽培。叶片椭圆状卵形，边缘有尖锐的单或重锯齿，边缘具芒刺，叶柄有腺体。花期3月底至4月初，晚于东京樱花；花3-6朵成伞房状总状花序，花序梗短；花两性，先叶开放，萼筒管状，带紫红色，萼片边缘有细齿，花瓣为重瓣，初时粉红色，后转白色。著名观赏植物，塞万提斯像东侧和中古史中心等地有栽培。

Deciduous tree. This variety of *C. serrulata* is native to Japan, and widely cultivated in gardens as an ornamental tree in China. It is planted near the statue of Cervantes on campus.

郁李

蔷薇科樱属
Cerasus japonica
(oriental bush cherry)
Rosaceae

落叶灌木，分布我国长江以北。小枝无毛。单叶互生，在芽内为对折状，叶片长卵形，先端渐尖，边缘具缺刻状重锯齿，叶柄短。花期4–5月，花两性，先叶开放，1–3朵簇生，萼片椭圆形，边缘有细锯齿，花瓣5；雄蕊多数。果期7–8月，核果，球形，熟时红色，核表面光滑。静园和南阁东面的花园中有栽培。

Deciduous shrub, native to North China, Japan, and Korea. This species is often cultivated for its flowers and medicinally used kernels. It is cultivated in Jing Yuan, and east of Nan Ge.

桃

蔷薇科桃属
Amygdalus persica (peach)
Rosaceae

落叶小乔木，原产于我国北方，有上千年的栽培历史，现在世界温带和亚热带广为种植，品种很多。典型的形态为小枝红褐色或褐绿色，单叶互生，在芽内为对折状，椭圆状披针形，先端具长尖，边缘有较密锯齿。花期4-5月，花两性，2朵并生，先叶开放，无柄，萼片平展，花瓣5，通常粉红色；雄蕊多数，花药紫色。果期6-8月，核果，卵球形，果核表面具沟和皱纹。著名果树，校园栽培有多个品种，以重瓣的碧桃（*A. persica* f. *duplex*）为多。未名湖东北岸的几株桃树与博雅塔遥相呼应，是早春的最"上镜"的植物之一。

Deciduous tree. Although native to North China, truly wild peaches no longer exist. It has been cultivated in China for several thousand years. Peaches are important, fruit-bearing and ornamental plants cultivated throughout temperate and subtropical zones with many cultivars. In Chinese culture, peach fruit is a symbol of longevity, and its seeds are used medicinally for reducing inflammation and allergies. Several cultivars are widely cultivated on campus, together with its cultivated double flower form (*A. persica* f. *duplex*). Several trees exist on the northeast side of Wei Ming Lake and are the "focal point" for taking photos in the spring.

这种古老的植物拥有那么多奇异的象征意义。因为《诗经》里的一句"投我以木桃，报之以琼瑶"就和香艳之情纠结起来，而桃花偏又开得那么艳丽逼人；与"逃"的谐音又让桃木成了道士手里杀鬼的法器，垂花门前挂着的辟邪的木符；蟠桃致长寿之说又让它成为仙界的草木。（王立刚）

榆叶梅

蔷薇科桃属
Amygdalus triloba
(flowering almond)
Rosaceae

落叶灌木，北方栽培多。单叶互生，在芽内为对折状，椭圆形至倒卵形，先端常3浅裂，边缘有重锯齿。花期3–4月，花两性，2朵并生，先叶开放，有短柄；花瓣粉红色，5枚或为重瓣；雄蕊20。果期5–6月，核果，近球形，被毛。常见观赏植物，未名湖边及文史楼、地学楼周边均有栽培，其他地方也多见。

Deciduous shrub, native to North, Northeast, and East China, Korea, and Russia. This species is widely cultivated in China. Single and double flower cultivars are planted as garden ornamentals. Several cultivars are planted west of the former Biology Building, around Wei Ming Lake, and elsewhere.

没有哪一种植物敢像它一般，在叶子未展时就开放如此浓密的花，总担心瘦弱的枝条承载不了沉甸甸的花串。看了它才知道"春意闹"有多闹，在推崇幽淡的文人那里，它必是俗的。但要抵挡它的活色和生气实在太难。（江都）

杏

蔷薇科杏属
Armeniaca vulgaris (apricot)
Rosaceae

落叶乔木，原产于亚洲西部，我国除华南外各地栽培。小枝褐色，有光泽。单叶互生，在芽内为席卷状，卵圆形，先端具尾尖，边缘具钝锯齿。花期4月，花两性，单生，先叶开放，萼片在花开后反折，花瓣5，白色或浅红色；雄蕊多数，花药黄色。果期6—7月，核果球形，果核平滑。著名果树，俄文楼西草地之西南区，以及民主楼前有栽培。

Deciduous tree. This species produces fruit known as apricots. Because of its long history of cultivation in China, it is difficult to know for certain whether they are really wild or escaped from cultivation, but probably the species originated in Central Asia. The seeds are also edible and used medicinally. Several trees are planted in front of the Democracy Building, and the Department of Philosophy.

在文艺作品中柔弱而美丽的杏花多与美女有关，但也时常挂着负面报道，如出轨、风流，甚至薄命。我们容易想起"一枝红杏出墙来"等等。《红楼梦》里探春对应于杏花，身份倒是不赖："相国栽培物，仙人种植花。"探春最后远走高飞了，传说中还成为王妃。比起袭人所对应的桃花，杏花要算好的了。（刘华杰）

桃、碧桃、山桃、杏、榆叶梅的区别

　　桃和碧桃的差别在于后者是重瓣花，较大。二者与山桃的区别是前二者的花萼片外侧有绒毛，后者萼片外侧无绒毛，另，前二者的叶为椭圆状披针形或长圆披针形，其最宽处约在中部，后者的叶为卵圆披针形，先端长渐尖，最宽处在中部以下。

　　杏和前三者不同处在于杏花萼片外侧无绒毛，萼片反折，这一点在其花瓣脱落后更为明显，又杏的叶宽大，呈卵圆形或近圆形。明显不同于前三者。

　　榆叶梅为灌木，叶片宽卵形，倒卵形，常三裂，边缘有重锯齿，花托（或称萼筒）广钟形，均不同于前四者。榆叶梅的花粉红色，有单瓣也有重瓣的。

玉兰

木兰科玉兰属
Yulania denudata (yulan magnolia, synonym *Magnolia denudata*)
Magnoliaceae

落叶乔木，原产于我国中部，现广为栽培。冬芽大，密被灰绿色或灰黄色绒毛。单叶互生，倒卵形，全缘；托叶膜质，早落后留下环状托叶痕。花期 4 月初，花两性，单生小枝顶端，先叶开放，大而美丽；花被 9 片，萼片与花瓣无明显区别，全为白色或稍带紫红色。果期 5–11 月，聚合蓇葖果，圆柱形。著名观赏植物，图书馆南门外有 2 株，经济研究中心院内有 1 株，校景亭西侧新近种了一棵，长势很好；陈守仁国际研究中心西侧可能是一个杂交种。

Deciduous tree, endemic to central and East China. This species has a long history of cultivation in China, and is widely cultivated in temperate parts of the world. Its white flowers were regarded as a symbol of purity in ancient China, and are one of the most frequent subjects for Chinese paintings. Several trees are planted south of the library and in Yan Nan Yuan.

我抬头看窗外,首先看到的就是那一棵玉兰花树,此时繁华久落,绿叶满枝。我仿佛听到在雨滴敲击下左右翻动的叶子正在那里悄声互相交谈:"伙计们!尽量张开嘴巴吮吸这贵如油的春雨吧!"我甚至看到这些绿叶在雨中跳起了华尔兹,舞姿优美整齐。我头顶上铁板的敲击声仿佛为它们的舞步伴奏。可惜我是一个舞盲,否则我也会破窗而出,同这些可爱的玉兰树叶共同蹁跹起舞。(季羡林)

紫玉兰

木兰科玉兰属
Yulania liliiflora (purple magnolia, synonym *Magnolia liliiflora*)
Magnoliaceae

又称辛夷。落叶灌木或小乔木，原产于我国中部和西南部，现全国广泛栽培。叶形与玉兰相似，具环状托叶痕。花期4月中旬，花单生小枝顶端，先叶开放或与叶同时开放；花被片9，明显分为3枚绿色的披针形萼片和6枚紫红色的披针形花瓣。果期5–7月，聚合蓇葖果，圆柱形。观赏植物，是另一个广为栽培的二乔玉兰（*Y.* × *soulangeana*）的亲本之一，这两个物种校园近期引种较多，陈守仁国际研究中心草地、红三楼北边、农园食堂北边，经济研究中心等地均有种植。

Deciduous shrub, endemic to central and Southwest China. This species is widely cultivated in China and Japan, and elsewhere in the world for its upright showy purple flowers. It is one of the parents of the hybrid magnolica *Y.* × *soulangeana* which has dropping flowers, also known as saucer magnolia. Several plants of purple magnolia and saucer magnolia have been introduced on campus.

拟南芥

十字花科鼠耳芥属
Arabidopsis thaliana
(mouse-ear cress)
Brassicaceae

又称鼠耳芥。一年生草本，我国西北及长江流域有野生。植株小，全株被毛。单叶，基生叶有柄，呈莲座状，叶片倒卵形或匙形；茎生叶无柄，披针形或线形。总状花序顶生；花两性，花瓣4，白色，成十字形；四强雄蕊。长角果细长，成熟时开裂。生长期45天左右。拟南芥为植物学研究中的模式植物之一，具各种突变体，在表型上有很大差异。生科院实验室栽培极多。

Short-lived annual herb. This species is native to Eurasia, and introduced to North America in the 19th century. It is the first flowering plant whose genome has been sequenced, and it is most widely used as the model organism for studies in plant genetics, development, physiology, biochemistry, and related fields. It is also a naturalized weed throughout much of the world. It is probably the most abundant plant species on campus – many laboratories cultivate this species for experimental purposes.

紫丁香

木犀科丁香属
Syringa oblata (early lilac)
Oleaceae

落叶灌木，原产于我国。枝条粗壮无毛。单叶对生，广卵形，基部心形或圆形，全缘，两面无毛。花期4月，圆锥花序侧生；花两性，极芳香，花冠紫色，高脚杯状，先端4裂，开展；雄蕊2。果期7–8月，蒴果，长圆形，2裂。著名观赏植物，校园多有栽培。还有花为白色的变型白丁香，校园也有栽培。

Deciduous shrub, native to central, North, and Northeast China, and Korea. This species is one of the most famous ornamental plants, and widely cultivated in China for its showy and fragrant flowers. It is also used as traditional Chinese medicine. It is commonly cultivated on campus along roadsides. Plants with white flowers have been recognized as *S. oblata* f. *alba*, which is also planted on campus.

城外校园里丁香更多。最好的是图书馆北面的丁香三角地,种有十数棵白丁香和紫丁香。月光下白的潇洒,紫的朦胧。还有淡淡的幽雅的甜香,非桂非兰,在夜色中也能让人分辨出,这是丁香。在我住了断续近三十年的斗室外,有三棵白丁香。每到春来,伏案时抬头便见檐前积雪。雪色映进窗来,香气直透毫端。(宗璞)

皱皮木瓜

蔷薇科木瓜属
Chaenomeles speciosa (flowering quince)
Rosaceae

又称贴梗海棠。落叶灌木，原产于我国，北京有栽培。枝条常具刺，小枝无毛。单叶互生，卵形，边缘具锯齿，齿尖有腺体，两面光滑；托叶肾形或椭圆形，边缘有尖锐重锯齿。花期3—5月，花两性，先叶开放，花瓣5，猩红色；雄蕊多数。果期9—10月，梨果，球形，黄色或黄绿色。观赏植物，南阁西南草坪、西门水池边，以及塞克勒考古博物馆东边有栽培。

Deciduous shrub, native to China. This species is often cultivated for its red or white flowers, and medicinally the fruit is used for treating arthritis, leg edema, and cramping in the calf muscle. It is cultivated near the pond at the West Gate, and by the Arthur M. Sackler Museum of Art and Archaeology.

贴梗海棠因为果实干后果皮会皱缩,所以也叫"皱皮木瓜"。尽管皱皮木瓜也能食用,但并非今天市面上卖的"木瓜",今日之木瓜是原产热带美洲的一种乔木的果实,准确的名字应该叫"番木瓜"。(刘夙)

蒲公英

菊科蒲公英属
Taraxacum mongolicum
(golian dandelion)
Asteraceae

多年生草本植物，我国广泛分布。植株具白色乳汁。单叶基生，莲座状，一般为大头羽裂，裂片三角形，全缘或有数齿。花果期3-6月，头状花序在花葶上顶生，总苞片数层，内层总苞片顶端具小角状突起；花全为舌状花，两性，鲜黄色，先端平截，5齿裂。瘦果倒披针形，有长喙，冠生于喙顶。常见杂草，校园多野生，凡草地上都可见。

Perennial herb, native to China, Japan, Korea, Mongolia, and Russia. This species is a widely distributed weed. The young plant is edible as a vegetable and the whole plant is used medicinally. It is very common on campus.

虽然田野上的草有千万种,但蒲公英是田野最富诗意的象征,吟诵它的童谣,世界上不知道有千万首。细细的花葶举起雪白的毛团,如同一朵朵等待绽放的云。孩童就像热爱自己的美梦一样热爱它们,并且祈祷自己的父母不要挖尽这些神奇的植物当作春炊中的野菜。当在燕园里和一朵蒲公英重逢,不妨以童年的心意祈祷每一颗诞生在小小云朵里的种子都会找到一小片泥土。(王立刚)

元宝枫

槭树科槭树属
Acer truncatum (Shandong maple)
Aceraceae

又称平基槭。落叶乔木，分布于我国长江以北地区。树皮灰褐色，深纵裂。单叶对生，通常掌状5深裂，裂片先端渐尖，叶基通常截形，两面均无毛，秋季转为红色。花期4-5月，伞房花序顶生；花小，杂性，黄绿色。果期9-10月，双翅果。红叶植物，校园栽培多，且多老树，如标本馆与老生物楼一段的马路北侧均有，湖心岛上的一些大树秋天叶红时，颇为壮观。

Deciduous tree, native to North and Northeast China, and Korea. This species is widely cultivated in temperate regions for its yellow and red ornamental foliage in the fall. There is a most beautiful one on the island in Wei Ming Lake.

北大之美一半在未名湖，而湖中之岛，如美人之目。秋天时元宝枫由绿而黄最后变成深红，仿佛美人眼睛上靓丽的蛾眉和花钿。湖心岛东南角岸边那株元宝枫因为正对着博雅塔，更是成了无数秋日登岛照片中必有的合影者。（赵瑞白）

落叶乔木，我国南北均有分布。雌雄异株，嫩枝及嫩叶含白色乳汁。单叶互生，卵形或广卵形，边缘有粗锯齿或深裂，为家蚕的饲料。花期5月，柔荑花序。果期6月，果序结果时肉质化，形成聚花果，黑紫色，通称桑葚，酸甜可食。西校门内水池东北岸及第一教学楼东南侧有老树为雄株，校史馆北面有雌株，甚高大，未名湖北山坡上也有不少由种子长出的小桑树。

Deciduous shrub or tree, endemic to central and North China. This species has a long history of cultivation in China, and the leaf is the feed for silkworms. The bark fiber is used for textiles and paper, the bark for medicine, and the fruit is edible. A male tree near the pond of the West Gate of campus is more than 300 years old.

桑

桑科桑属
Morus alba (white mulberry)
Moraceae

朗润园湖岸上有两株古桑,或许是因为岸基松动,整个儿倾向湖中心,堪堪掠过水面。大概桑是甘于奉献的,没有纤纤素手来采去叶子喂蚕,只得将红红紫紫的葚实结了满枝,招来成群成队贪吃的鸟儿。(蔡乐)

构树

桑科构属
Broussonetia papyrifera (paper mulberry)
Moraceae

又称楮树。落叶乔木，分布我国南北各省。雌雄异株。小枝粗壮，密生绒毛。单叶互生，不裂或具不规则的 3–5 深裂，叶缘具粗锯齿，两面被毛。花期 5–6 月，雄花组成柔荑花序，雌花组成球形头状花序。果期 9–10 月，聚花果，球形，成熟时肉质，橘红色，可以食用。其茎皮纤维为古代常用的造纸原料。校园有野生植株，也有老树。

Deciduous tree, native to eastern Asia. This is a fast-growing species. The bark is composed of very strong fibres, and can be used for making high-quality paper. The wood is used for furniture, and the leaves, fruit, and bark are used medicinally. When introduced to non-native areas it can quickly disrupt the native habitat, becoming a highly invasive species, in Latin America, the United States, and South Asia. It is sporadically seen on campus.

构树虽然也能长到十几米高,但因其烂芯故难成材。虽然不能做栋梁,但构树之用却很广,单从其俗名如肥猪树、纱树、谷浆树云云,就可见其用处之广,台湾民间采其叶饲鹿故名其曰"鹿仔树"。树皮可造纸,果实、树汁可入药,枝干又可做薪材。《酉阳杂俎》上说:"构,田废久必生。"可见构树不择地而生,适应性之强悍。(刘华杰)

香茶藨子

落叶灌木，原产于美国，北京有栽培。枝不具刺，幼枝密被白色柔毛。单叶互生，3 裂，先端具粗钝齿牙，基部截形，下面有棕褐色锈斑。花期 5 月，总状花序下垂；花两性，黄色，萼裂片反卷，花瓣 5，黄色，长为萼片一半。果期 6-9 月，浆果，近球形，黑色。观赏植物，南阁东边草坪有栽培。

Deciduous shrub, native to North America. This species is cultivated in the northern part of China for its golden yellow and fragrant flowers. It is planted east of Nan Ge.

虎耳草科茶藨子属
Ribes odoratum
(golden currant or Missouri currant)
Saxifragaceae

点地梅

报春花科点地梅属
Androsace umbellata
(umbellate rock jasmine)
Primulaceae

一年生草本，分布于我国南北各省区。全株被长柔毛。单叶，基生叶丛生，叶近圆形，先端钝圆，边缘有多数三角状钝牙齿。花期4–5月，花葶通常数条，由基部叶腋抽出，伞形花序；花萼杯状，5深裂几达基部，花冠通常白色，裂片5，喉部黄色。果期6月，蒴果，近球形，成熟后5瓣裂。未名湖西北岸山坡草丛和校景亭西侧草地多见。

Annual or biennial herb, native to Asia and very common on open grassy areas or roadsides. One of its Chinese names is based on its medicinal use to cure sour throats – throat herb. It is often seen among grasses near Hong Hu (Red Lake) on the western part of the campus.

走在北大校园，一低头，草丛或路边白色的星星点点随意却不凌乱，那便是点地梅了，真是贴切极了的名字。若是俯身细看，精致的五瓣小花，中心是一点鲜艳的鹅黄，在姹紫嫣红的春夏之际低调地美丽着。（吴岚）

黄栌

漆树科黄栌属
Cotinus coggygria
(Eurasian smoke tree)
Anacardiaceae

又称红叶。落叶灌木或乔木，分布于我国北方。小枝紫褐色，有刺鼻气味的白色乳汁。单叶互生，圆形或倒卵形，全缘，秋季转为红色。花期4月，大型圆锥花序顶生；花杂性。果期6月，果序具多数宿存的不孕花的细长花梗，紫绿色羽毛状。为著名的红叶植物，是秋天"香山红叶"的主要树种。校园栽培的均是该物种的一个变种：灰毛黄栌。

Deciduous shrub, native to a large area from southern Europe, central Asia and the Himalaya to northern China. This is an important ornamental plant for its colorful foliage in the fall. This is the dominant tree species on Fragrant Mountain in Beijing. One variety of this species, var. *cinerea*, is commonly cultivated on the hills on campus.

俄文楼……北面小山下有几树黄栌，是涂抹秋色的能手。……黄栌一年一度焕彩蒸霞，使这一带的秋意如醇酒，如一曲辉煌的钢琴协奏曲。（宗璞）

黄杨

黄杨科黄杨属
Buxus sinica (Chinese boxwood)
Buxaceae

常绿灌木，原产于我国。雌雄同株。树皮灰白色，茎四棱形。单叶对生，叶形较小，革质，正面呈深绿色，背面为浅绿色，全缘，侧脉明显。花期4月，花簇生，浅黄色，无花瓣，雄花萼片4，雌花萼片6。果期6–7月，蒴果，近球形，具3枚宿存花柱。常见绿篱植物，校园栽培很多，如俄文楼西侧草地北部即是。

Evergreen shrub or small tree, native to China. This species is most commonly used as hedge and bonsai material. It is planted in front of the Arthur M. Sackler Museum of Art and Archaeology on the west part of campus.

华山松

松科松属
Pinus armandii
(Chinese white pine)
Pinaceae

常绿乔木，我国华北及西南皆有分布，现广泛栽培。雌雄同株。一年生小枝绿色，无毛。叶针状，5 针一束。花期 4–5 月。次年 9–10 月种子成熟，球果大，圆锥状卵圆形；种子无翅，可以食用。为园林树种，校园图书馆东广场的西北角、三角地、南阁附近等地有栽培。

Evergreen tree, native to North and Southwest China. This species is important in forestry plantations in some parts of China, and also grown as an ornamental tree in parks and large gardens. It was introduced to Europe and North America. Its seeds are edible as pine nuts, and the wood is used for general building purposes. It is planted near the Library, Nan Ge, and Triangle Area on campus.

白皮松

松科松属
Pinus bungeana (lacebark pine)
Pinaceae

常绿乔木，原产于我国华北以及西北。雌雄同株。有时几个树干簇生似灌木状，树皮灰绿色，呈不规则片状脱落，脱落后露出白色内皮。叶针状，3 针一束。花期 5 月。次年 10 月种子成熟，球果单生，卵圆形；种子有翅。为优美的园林树种，是校园的常见种，临湖轩南边的草地上有两株老树，树龄超过 300 岁。

Evergreen tree. It is native to China, and widely cultivated as an ornamental plant in East Asian temples and classical gardens where it symbolizes longevity. Characteristics include: smooth, grey-green bark, which gradually peals away to reveal patches of pale yellow, which turn olive-brown, red and purple on exposure to light. It is a commonly seen species on campus and some are older than 300 years, such as two trees southeast of Lin Hu Xuan.

临湖轩前面的两株白皮松，是很壮观的。它们有石砌的底座，显得格外尊贵。树身挺直，树皮呈灰白色。北边的一株在根处便分杈，两条树干相并相依，似可谓之连理。南边的一株树身粗壮，在高处分杈。两树的枝叶都比较收拢，树顶不太大，好像三位高大而瘦削的老人，因为饱经沧桑，只有沉默。（宗璞）

银杏

银杏科银杏属
Ginkgo biloba (ginkgo tree)
Ginkgoaceae

落叶乔木，原产于我国（可能在浙江天目山），现世界各地广泛栽培。雄雌异株。雄树枝条一般斜上生长，雌树枝条则较为开展。叶扇形，先端常2裂；叶脉为种子植物中罕见的二叉分支。花期4–5月。当年10月种子成熟，核果状；外种皮为肉质，有臭味，中种皮骨质；种仁通称白果，可食用。为著名的园林树种，校园很多，尤以图书馆东广场周围、学生宿舍和未名湖北岸为多，西门内有老树，其树叶秋天变成金黄色，成为校园著名的风景。

Deciduous tree. This species may be native to northwest Zhejiang (Tianmu Shan), and it is widely cultivated in China and other countries in the world probably for more than 3000 years. It provides shade and is tolerant of a wide range of climatic conditions, and it is sacred to Buddhists and often planted near temples. The wood is used for furniture, the leaves are used medicinally and used for pesticides, the roots are used as a cure for leucorrhea, and the seeds are edible. It is a common tree species on campus with many over 100 years old. The scenery with its bright yellow foliage is on many pictures describing PKU campus.

西门内是北大最壮观的园林布局。轩楼朱阁,飞檐嵯峨。如果不是这颗银杏生得天庭饱满,地阁方圆,枝雄干壮,外秀慧中,怎能压住这里的氛围。清秋气穆,灿然的落英和白果,陨堕如雨,仰首其下,觉得她占满整个天空,并且如同天空一样有尊严。(王立刚)

水杉

杉科水杉属
Metasequoia glyptostroboides
(dawn redwood)
Taxodiaceae

落叶高大乔木，原产于我国中部，现世界广泛栽培。雌雄同株。主干挺直；小枝分长枝和脱落性的短枝，对生，下垂。叶片条形，交互对生。花期4月。当年10月种子成熟，球果近球形，下垂。水杉为速生性树种，树形优美，常用作园林树；现在世界上所有栽培水杉树的谱系均能追溯到最早发现的那些"母树"。老生物楼门外东西侧各有一株老树，此外在老生物楼后小楼前、南阁东面的花园和二院南面也有栽培。

Deciduous tree, endemic to China. The *Metasequoia* was first described as a fossil genus from the Mesozoic Era in 1941. It was discovered in Wan County (presently, Lichuan County, Hubei) in 1944, and finally described as a new living species in 1948 by the Chinese botanists Cheng Wan-Chun and Hu Hsien-Hsu. After that, the seeds were distributed to botanical gardens and universities all over the world. Although it enjoys world-wide cultivation, it is regarded as a rare species as a native plant because the original habitat is now largely under rice cultivation. Several trees are found north of the former Biology Building, and east of Nan Ge, and in front of Er Yuan.

1941年，日本古植物学家三木茂发现一种产自日本中生代地层的奇特杉科植物化石，当时所有人都以为这种古老的植物已经灭绝了。然而就在两年后，中国学者王战在四川万县磨刀溪（今属湖北利川）发现了一棵奇特的杉树，著名植物分类学家胡先骕确定这种杉树就是先前人们认为已经灭绝的"变形红杉"。像这样先发现化石、再发现现生种的事例，在分类学史上是极为罕见的。崎岖闭塞的地形少人砍伐，而重峦叠嶂为冰期中的植物提供了良好的避难所，这才存续了中国很多珍稀的孑遗植物。（刘夙）

白杆

松科云杉属
Picea meyeri (Meyer spruce)
Pinaceae

常绿乔木，原产于我国华北地区，现各地广泛栽培。雌雄同株。一年生小枝淡黄色或黄褐色，常有毛，上面生有众多钉状的叶枕。叶生于叶枕上，锥形，横切面菱形。花期4–5月，雌球花单生枝顶，雄球花单生叶腋，皆下垂。当年9–10月种子成熟，球果下垂。为材用树种和园林树种，校园有多株老树，办公楼西侧，老生物楼东北侧，三角地花坛中均有种植。

Evergreen tree. This is the dominant species in the mixed forests of coniferous and broad-leafed trees endemic to North China. It is also cultivated for reforestation and as an ornamental. It was introduced to eastern USA to replace blue spruce. Several trees are found sporadically on campus.

北阁西边以及办公楼西侧有白杆，此树就笔者记忆，五十年代初就有，几十年来似乎未长高多少，此树本来就是生长慢的。可见其年岁恐也在百年以上了。（汪劲武）

油松

松科松属
Pinus tabuliformis
(Chinese red pine)
Pinaceae

常绿乔木，我国长江以北广泛分布。雌雄同株。树冠平展似桌形，一年生小枝淡褐色，无毛。叶针状，2针一束。花期4—5月。次年9—10月种子成熟，球果卵球形；种子有翅。为北方重要园林树种，也是北方山区重要绿化和水土保持树种。校园极多，湖心岛亭和静园有树龄超过三百岁的老树。

Evergreen tree, native to central, North, and Northeast China, and Korea. This species is widely cultivated in China as an ornamental for its table-shaped canopy. It is also planted for timber which is used for construction, poles, ship building, and furniture. The trunk can be used as a source of resin, the bark for tannin, and the needles for medicine. Many trees are found on campus, with a few over 300 years old (the ones on lake island and Jing Yuan), and several others are over 100 years old.

岛亭相伴300多年树龄的古老油松，树冠浑圆如盖，主干欹斜不倒，枝干如群龙漫舞，逸气横生，为湖心岛增添了几分蓬莱仙境的意趣。（谢凝高）

圆柏

柏科刺柏属
Juniperus chinensis (Chinese juniper)
Cupressaceae

又称桧柏。常绿乔木，原产于我国、日本等地，现广泛栽培。雌雄异株。不同品种的株形各异。叶通常有刺形（3叶轮生）和鳞形两种，但在不同品种的植株上数目不同，有的品种可全为刺叶或全为鳞叶。花期4月。当年10月种子成熟，球果近球形，蓝绿色，种鳞肉质，不开裂。为我国著名园林树种，品种很多。校园栽培较多，不少树龄超过了一百年，如第一教学楼前的老树。

Evergreen shrub or tree, native to China, Japan, Korea, and East Russia. This is a widely cultivated species in gardens and parks, with numerous cultivars. Many trees around the Wei Ming Lake area and by some buildings are over 100 years old.

侧柏

柏科侧柏属
Platycladus orientalis
(Chinese arborvitae)
Cupressaceae

常绿乔木，原产于我国北部，现全国广泛栽培。雌雄同株。枝条开展，小枝扁平，排列成复叶状。叶全为鳞片状，交互对生。花期4–5月，球花生于枝顶。当年10月种子成熟，球果卵球形，成熟时种鳞成木质，开裂。为重要园林树种，北京市市树之一（另一种是槐），校园栽培很多，第一教学楼前有老树。也是重要的绿化和水土保持树种，但生长极为缓慢。

Evergreen tree, native to Northwest China and widely naturalized elsewhere in Asia. The common name "arborvitae" is a Latin word, meaning "tree of life", and is based on its association with longevity in Chinese Buddhist thought. Some trees planted around Buddhist temples in China are more than 1000 years old. It is commonly planted on campus, with some more than 100 years old.

一教门前有两株长得并不高大却已年逾百岁的老侧柏，每到秋季球果成熟之时，一颗颗咧着小口的圆球掉落满地，总引得鸟儿争相啄食。在学习疲倦时透过窗户眺望这两株古树，它们俩始终像两个谦卑而温和的老人守护着教学楼。（张慧婷）

胡桃

胡桃科胡桃属
Juglans regia (common walnut or Persian walnut)
Juglandaceae

俗称核桃。落叶乔木，我国南北各省广泛栽培。雌雄同株。老树树皮纵裂，呈深浅色相间的纵纹。奇数羽状复叶互生，小叶全缘，顶端小叶最大，光滑无毛。花期4−5月，雄花组成柔荑花序，下垂，雌花组成穗状花序，通常1−3朵花。果期9−10月，果序短，具1−3个果；果实为核果，近球形，无毛。著名果树，学生宿舍楼前栽培较多。

Deciduous trees, native to central Asia. This species has a very long history of cultivation in China and elsewhere in the world, mostly for its edible, oily nuts and hard, fine grained wood. It is cultivated in student dormitory areas on campus.

男生抱起他热恋的女生摘取核桃，或者拿着树枝击落头顶上尚有青色的核桃，这是燕园里最美丽的景致之一了。一粒核桃要在哪间宿舍的窗沿上晒多久？要在哪对情侣的手里握多久？那一点点核仁的滋味要在他们的心中留多久？（王立刚）

七叶树

七叶树科七叶树属
Aesculus chinensis
(Chinese horse chestnut)
Hippocastanaceae

落叶高大乔木，原产于我国华北。掌状复叶大，对生，有长柄，小叶 5–7。花期 4–5 月，圆锥花序，花杂性，白色。果期 10 月，果实球形或倒卵圆形。为著名园林树种，北方佛教寺庙多栽培，用以代替娑罗双树。校园栽培少，老生物楼门外东侧有一株大树，高达房檐以上，干直径约 40 厘米，树龄比楼的年龄大；俄文楼西北侧也有一株成年树，花期时甚美。

Deciduous tree, native to China. Its beautiful shape and dense flowers make it a famous ornamental tree in gardens and parks. It is often planted in temples in northern China. After processing the seed is edible and also used medicinally. Its wood is used for carving and furniture. The tree in front of the former Biology Building is much older than the building.

美国红梣

木犀科梣属
Fraxinus pennsylvanica (green ash)
Oleaceae

落叶乔木，原产于北美。雌雄异株。奇数羽状复叶对生，小叶 7-9，长圆状披针形，边缘具不明显钝锯齿。花期 4-5 月，圆锥花序，生于二年生枝上，与叶同放。果期 8-9 月，翅果，倒披针形，果翅下延达果体 1/2 以上。为良好的行道树种，校园栽培的梣属树木多为本种。

Deciduous tree, native to North America. This species is introduced to the northern part of China as shade trees along the sidewalks or ornamental trees in gardens and parks. It is a common tree species on campus.

栓皮栎

壳斗科栎属
Quercus variabilis
(Chinese cork oak)
Fagaceae

落叶乔木，分布于我国南北各省。雌雄同株。树皮黑褐色，条状纵裂，木栓层发达，厚可达 10 厘米，为软木塞的原料。单叶互生，长圆状披针形，背面有白色星状毛，叶缘具芒刺状锯齿。花期 4–5 月，雄花成下垂的柔荑花序，雌花单生或几个聚生。果期 9–10 月，坚果（通称橡实）外有杯状的壳斗，包围坚果 2/3 以上，壳斗上的鳞片锥形，向外弯曲。校园临湖轩及西北山丘上有多株较老的树。

Deciduous tree, native to China, Japan, and Korea. It is a common tree species in the mixed evergreen and deciduous forests. It is sometimes cultivated for cork production or as an ornamental in China. There are several big trees on the hills south of Wei Ming Lake.

斑种草

紫草科斑种草属
Bothriospermum chinense
(China spotseed)
Boraginaceae

一年生草本，分布于我国北方。植物密被刚毛。单叶互生，长圆形，边缘皱波状，两面有短糙毛。花期4-6月，花序具叶状苞片；花小，花萼裂片5，有毛，花冠淡蓝色，5裂，喉部具鳞片状附属物。果期6-8月，小坚果4，肾形。为常见杂草，校园路边、绿篱下、草地上、山坡上极多见。

Annual or rarely biennial herb. It is endemic to North China. The whole plant is used in traditional Chinese medicine to reduce inflammation. It is a common weed on campus.

附地菜

紫草科附地菜属
Trigonotis peduncularis
(pedunculate trigonotis)
Boraginaceae

一年生矮小草本，我国南北各省有分布。茎纤细，具平伏细毛。单叶互生，匙形，两面均具平伏粗毛；下部叶具短柄，上部叶无柄。花期4-5月，花序顶生；花小，通常生于花序的一侧，花萼5裂，花冠蓝色，5裂，喉部具鳞片状附属物。果期7-8月，小坚果4，三角状四边形。为常见杂草，校园极多，路边绿篱下，草地上多见。

Annual or biennial herb, distributed in temperate Asia, and East Europe. The whole plant is used as herbal medicine for relieving stomach-ache. It is a common weed on campus.

抱茎小苦荬

菊科小苦荬属
Ixeridium sonchifolium
(amplexicaul ixeridium)
Asteraceae

又称抱茎苦荬菜。二年生草本，我国北方常见。植株具白色乳汁，茎多单生，圆柱形。基生叶莲座状，叶片倒卵状披针形，边缘常为羽状齿裂或深裂；茎生叶长卵状披针形，边缘羽状齿裂或深裂，基部两侧膨大，呈耳状抱茎。花果期4-7月，头状花序小，排列成伞房状；花全为舌状花，花冠淡黄色。果实为瘦果，具喙，顶生白色冠毛。为常见杂草，校园草地、路边常见。

Perennial herb, native to central and North China, and Japan and Korea. This species is a common weed on roadsides on mountain slopes, plains, floodplains, and rocky terrain. It is used as herbal medicine, and as animal feed. It is a common weed on campus.

抱茎小苦荬、中华小苦荬的区别

抱茎小苦荬的茎生叶基部膨大并完全抱茎，中华小苦荬的茎生叶仅1-2个，基部微抱茎但不膨大。

中华小苦荬

菊科小苦荬属
Ixeridium chinense
(Chinese ixeridium)
Asteraceae

又称中华苦荬、苦菜。多年生草本，分布于我国东部、南部和北部地区。植株有白色乳汁。基生叶莲座状，条状披针形，全缘或不规则羽裂；茎生叶抱茎。花果期4-7月，头状花序较抱茎小苦荬大，排列成伞房状。花全为舌状花，20个左右，淡黄色或白色。果实为瘦果，红棕色，冠毛白色。为常见杂草，校园很多，凡路边绿篱下、大草地中皆有。可作野菜食用。

Perennial herb, native to North, central, and East China, and Japan, Korea and Russia. This is a weedy species. The young plants are edible. The whole plant is used medicinally. It is very common on campus.

每年四月开始，中华小苦荬的小黄花就会开遍燕园的每个角落。若是有心，你会惊异于它们的纤细、轻灵、柔弱和精致，每一朵都如一件精巧的艺术品。它们原是农村里常见的饲料，谁知却有这样的柔美、灵秀。（蔡乐）

多年生草本，分布于我国长江以北各省。有发达的地下茎，鲜时黄色；植株密被灰白色柔毛和腺毛。单叶，通常基生，卵形至长椭圆形，上面绿色，下面略带紫色，边缘具不规则圆齿。花果期4-6月，总状花序顶生。花两性，略呈二唇形，花萼、花冠密被毛，花冠外面紫红色，内面黄色带紫斑。果实为蒴果，卵形；种子多数，表面有蜂窝状纹。著名药用植物，校园常见野生，红一楼北停车场路北的草丛中每年春季可见大量成簇开花的地黄。

Perennial herb, native to North and East China. This species is widely cultivated as a traditional Chinese medicine, and has efficacy in the scavenging of free radicals. It is a common weed on campus.

地黄

玄参科地黄属
Rehmannia glutinosa
(adhesive rehmannia)
Scrophulariaceae

中医中的常药却并不显眼。初夏时莲座叶间向上挺出一条花葶，上面挂一串低垂的花苞。有光的时候，细细的白色绒毛反射阳光，显得有些虚幻，又仿佛在逗弄着光影。而当那长长的花瓣终于在尾梢开展，花筒深处那一抹殷红隐隐绰绰地现出来，小小的植株竟一下子显出几分诱惑。地黄真正的宝藏在根里。古人以其根染衣，色黄而不褪，因而得名。（蔡乐）

牡丹

芍药科芍药属
Paeonia suffruticosa (tree peony)
Paeoniaceae

落叶灌木，原产于陕西，北京多有栽培。分枝粗而短。叶互生，二回三出复叶，顶端小叶3裂。花期5–6月，花单生枝顶，大而美丽，萼片5，绿色，花瓣5或常为重瓣，玫瑰色、红紫色、粉红色至白色；雄蕊多数。果期7–9月，蓇葖果，密生黄褐色硬毛。为著名观赏花卉，北大教育基金会东边花坛中有栽种，每年春天花开得特别艳丽。近年洛阳校友赠给北大一批牡丹已种植在静园、老地学楼西边、北大教育基金会院内。

Deciduous shrub, endemic to North China. It has been cultivated in China for hundreds of years and in gardens around the world, resulting in many varieties. Its bark is a famous Chinese traditional medicine "dan-pi". This species is also an important symbol in Chinese culture, representing happiness, prosperousness, richness and nobility, purity, and beauty. Several cultivars donated to PKU by the alumni from Luo-yang were planted in Jing Yuan and other parts of campus.

向来富贵的牡丹在这里如铅华洗尽，栖身于寻常百草木石之间。它的花期随着酷热一起到来，于是失去了人们的关注。但识得它的人，打眼间就会认出它繁复富丽的花瓣，即便蓬生于乱草间，也是"精华欲掩料应难"的。（蔡乐）

芍药

多年生草本，原产于我国北方，北京多有栽培。根粗壮，黑褐色。茎下部叶为二回三出复叶，上部叶为三出复叶，互生，小叶狭卵形、披针形或椭圆形，顶端小叶不裂。花期5-6月，花两性，数朵生枝顶或叶腋，萼片4，花瓣9-13，白色或粉红色；雄蕊多数。果期9月，蓇葖果，顶端具喙。为著名观赏花卉，静园内及校园多处有栽培。

Perennial Herb, native to North and Northeast China, Japan, Korea, Mongolia, and Far East Russia. This species is widely cultivated for its many double-flowered cultivars as ornamentals in the world. It is also used as a medicinal herb in traditional Chinese medicine. It is distinguished from the tree peony by its herbaceous habit. Several cultivars are planted on campus.

芍药科芍药属
Paeonia lactiflora (Chinese peony)
Paeoniaceae

它的艳丽丝毫不逊于它的姊妹牡丹，但却有"将离草"这么一个忧伤而美丽的名字。更因姜白石传唱千古的"念桥边红药，年年知为谁生"，浸足了乱离感伤的情调。（张慧婷）

巴天酸模

蓼科酸模属
Rumex patientia
(garden patience or patient dock)
Polygonaceae

多年生草本，我国北方广泛分布。单叶，基生叶大，长圆状披针形，长可达 15–30 厘米，叶缘波状；茎生叶互生，小而无柄；托叶鞘膜质。花期 5–8 月，圆锥花序顶生，花两性，花被片 6，排成内外两轮。果期 6–9 月，瘦果，三棱形，褐色，包于宿存、增大的内轮花被内；3 枚内轮花被均全缘，通常仅 1 枚具瘤状突起（易被误认为果实）。老生物楼大门外西侧常有生长。

Perennial herb, usually found along roadsides, ditches, and farm lands. The roots have been used in traditional Chinese medicine. It is native to northern temperate zones of Europe and Asia, and also introduced and naturalized in North America and other parts of the world. It is often seen along roadsides on campus.

日本小檗

落叶小灌木，原产于日本，现我国广泛栽培。茎具不分岔的刺。单叶互生，匙形或倒卵形，基部下延成短柄，全缘，上面黄绿色，下面灰白色，或均为紫红色，两面无毛。花期4-5月，花两性，单生或2-3朵组成伞形花序；花瓣黄色。果期7-9月，浆果，红色。栽培品种"紫叶小檗"在校园未名湖周围多有栽培，为北方常用的红叶绿篱植物。

Deciduous shrub, native to Japan. This is one of the most widely cultivated species of *Berberis*. It is commonly cultivated as an ornamental in China with many cultivars. Recently it was recognized as an invasive species in parts of the eastern United States. The cultivar with purplish leaves (*B. thunbergii* "Atropurpurea") is cultivated around Wei Ming Lake.

小檗科小檗属
Berberis thunbergii (Japanese barberry or Thunberg's barberry)
Berberidaceae

楸

紫葳科梓属
Catalpa bungei
(Chinese catalpa)
Bignoniaceae

落叶乔木，分布于我国长江流域。小枝灰绿色，无毛。单叶对生，三角状卵形，全缘，有时基部具1-4对齿，背面光滑。花期4-5月，总状花序成伞房状排列，顶生；花两性，较梓为大，花冠白色，二唇形，内有紫色斑点。果期6-9月，蒴果，长圆柱形。老生物楼东北侧、校景亭和邱德拔体育馆西侧都有栽培。

Deciduous tree, endemic to North and East China. This species is cultivated for many purposes: high quality timber, ornamental, and medicine. A big tree over 100 years old is found near Khoo Tech Puat Gymnasium on southeastern campus, and several smaller trees are found near Wei Ming Lake and the former Biology Building.

棣棠花

蔷薇科棣棠花属
Kerria japonica
(Japanese yellow rose)
Rosaceae

落叶灌木，分布于我国，北京有栽培。小枝绿色，有棱。叶卵形或三角状卵形，边缘有重锯齿。花期为春季的4-5月及秋季的9-10月，花两性，单生于侧枝顶端，萼筒扁平，花瓣5或为重瓣，黄色；雄蕊多数。果期7-9月，瘦果，萼片宿存。观赏植物，未名湖畔及校园其他地点有多种单瓣和重瓣的品种。

Deciduous shrub, native to China and Japan. This species is an ornamental plant in gardens and the double flower cultivar "Pleniflora" is very popular. Its flowers are used medicinally. Cultivar with double flowers is more commonly planted on campus.

黄刺玫

蔷薇科蔷薇属
Rosa xanthina
(northeastern Chinese rose)
Rosaceae

落叶灌木，原产于我国北部，北京栽培普遍。小枝有散生硬直皮刺。羽状复叶互生，小叶 7–13，宽卵形，边缘有钝锯齿，叶轴与叶柄均疏生柔毛和皮刺。花期 5–7 月，花两性，单生，较小，萼筒光滑，花瓣 5，黄色，倒卵形，重瓣或近重瓣，也有单瓣的；雄蕊多数。果期 7–9 月，蔷薇果，近球形，萼片宿存。观赏植物，校园栽培较多，未名湖边、学生宿舍区均有。

Deciduous shrub, native to North, Northeast and Northwest China. It is also frequently cultivated as ornamentals in gardens and parks, especially its double or semidouble flower form. It is commonly seen around Wei Ming Lake, student dormitories, and other parts of campus.

洋槐

豆科刺槐属
Robinia pseudoacacia (black locust, or false acacia)
Fabaceae

落叶乔木，原产于北美，我国长江以北栽培普遍。树皮灰褐色，浅至深纵裂。奇数羽状复叶互生，小叶7-25，椭圆形；托叶刺状，有时无托叶。花期5-6月，总状花序腋生，下垂；花两性，蝶形，白色，具香气；雄蕊10，2体(9枚合生，1枚离生)。果期7-9月，荚果，扁平带状，赤褐色。校园栽培甚多，开花季节香气宜人；俄文楼东北侧、办公楼南侧均有老树。

Deciduous tree, native to the southeastern United States. This species has been widely planted and naturalized in China and other temperate regions. The flowers have a pleasant fragrance and are the nectar source for honeybees. It is commonly seen on the western part of campus, and some of them are over 100 years old.

自从移家朗润园……就能看到成片的洋槐，满树繁花，闪着银光；花朵缀满高树枝头，开上去，开上去，一直开到高空，让我立刻想到新疆天池上看到的白皑皑的万古雪峰。（季羡林）

紫荆

豆科紫荆属
Cercis chinensis (Chinese redbud)
Fabaceae

落叶灌木，我国分布广泛，北京有栽培。单叶互生，叶全缘，基部心形，掌状脉。花期4-5月，花两性，于老干上簇生或成总状花序，先叶开放或与叶同时开放，两侧对称，花瓣5，紫色，上面3枚较小；雄蕊10，分离。果期8-9月，荚果，扁平，狭长椭圆形。观赏植物，一教大门东侧和学生宿舍区皆有栽培。

Deciduous shrub, endemic to central, North, and East China. The flowers of this species bloom on the stems before the leaves grow out in the spring. This species was a symbol of brotherhood or family tie in ancient China, and cited in some famous poems. It is commonly cultivated on campus.

人们熟知的香港区花"紫荆"其实是"洋紫荆",正式名称是"红花羊蹄甲"。北大校园里栽培的紫荆才是真正的紫荆,虽然花不如洋紫荆大,花期也只有春季的短短几个星期,却有更悠久的文化内涵,一千多年来一直被视为兄弟亲情的象征。杜甫《得舍弟消息》头两句就是:"风吹紫荆树,色与春庭暮。"(刘夙)

大花野豌豆

豆科野豌豆属
Vicia bungei (Bunge vetch)
Fabaceae

一至二年生草本，分布于我国北方。茎细弱，4棱。羽状复叶互生，具短叶柄，小叶6-10，长圆形或狭倒卵状长圆形，有小突尖，顶端小叶特化为分枝卷须，托叶半边箭形，具锐齿。花期4-5月，总状花序腋生，比叶长；花两性，蝶形，蓝紫色。果期6-7月，荚果，稍扁平，长圆形，含3-8粒种子。俄文楼南侧草地、西侧草地均有生长。

Annual herb, native to China and Korea. The plant could be used as animal feed. It grows on lawns on western campus.

落叶攀缘缠绕性大藤本，北京栽培多。奇数羽状复叶互生，小叶7-13，卵状椭圆形，全缘。花期4-5月，总状花序侧生，下垂，总花梗、小花梗及花萼密被柔毛；花两性，紫色或深紫色，雄蕊10，2体。果期8-9月，荚果，扁圆条形，密被白色绒毛，开裂后果皮呈螺旋状卷曲。著名观赏植物，静园一至三院内有老株。

Deciduous liana, native to North and East China, and Japan. This species is cultivated extensively in gardens and parks as a climbing ornamental. The whole plant is used medicinally. Its beautiful purple flowers and long life span make it a favorable garden plant, and favorable subject for Chinese paintings. Several old plants around Jing Yuan are very attractive when they are blooming.

紫藤

豆科紫藤属
Wisteria sinensis
(Chinese wisteria)
Fabaceae

从未见过开得这样盛的藤萝，只见一片辉煌的淡紫色，像一条瀑布，从空中垂下，不见其发端，也不见其终极，……我只是伫立凝望……它带走了这些时一直压在我心上的关于生死的疑惑，关于疾病的痛楚。我浸在这繁密的花朵的光辉中，别的一切暂时都不存在，有的只是精神的宁静和生的喜悦。（宗璞）

虞美人

罂粟科罂粟属
Papaver rhoeas
(field poppy, or red poppy)
Papaveraceae

一年生草本，原产于欧洲，现在世界广泛栽培。全株被开展粗毛，有白色乳汁。单叶互生羽状深裂或全裂，边缘有不规则锯齿。花期5-8月，花两性，单生，具长梗，花蕾时下垂；花瓣4，紫红色、红色至白色；雄蕊多数。果期6-9月，蒴果，近球形，光滑，成熟时孔裂；种子多数，细小。为常见观赏花卉，校园花坛（如电教楼西边、哲学楼东边的花坛）常有栽培。

Annual herb, native to North Africa, Southwest of Asia, and Europe. This species is widely cultivated in China as an ornamental with many cultivars of different floral colors and double flowers. It is commonly cultivated on campus.

若知道这个名字，人们在看到它时很难不想到那个凄美的乌江夜晚。它的惊艳让人感到不安，所谓烈酒最香，毒花最美。果然，它与罂粟是近亲。然而果实里有成瘾的汁液与花朵何干呢？（江都）

鸢尾

鸢尾科鸢尾属
Iris tectorum (wall iris)
Iridaceae

多年生宿根草本，原产于我国中部。单叶基生，渐尖状剑形，质薄，淡绿色，呈 2 纵列交互排列，基部互相套叠。花期 5-6 月，总状花序；花冠蓝紫色或紫白色，外轮 3 枚花被片较大，圆形，下垂，内轮 3 枚花被片较小，倒卵形，中央有一行鸡冠状白色带紫纹的突起。果期 6-8 月，蒴果，长椭圆形。观赏植物，未名湖边、红湖西岸有栽培。

Perennial herb, native to China, Japan, and Korea. This species has a long history of cultivation in China, and other parts of the world as an ornamental in gardens and parks. It is also used medicinally. It is commonly planted on campus.

鸢尾花柱的3个分支,都扩大成宽扁的花瓣状,顶端还要分叉,形似鹰(鸢)尾。有趣的是,在中国和西方,风筝的尾巴往往也都做成分叉状,所以风筝在汉语中又叫"纸鸢",而英语中的风筝(kite)干脆就和鸢是同一个词。据说风筝起源于中国(也有说法是起源于太平洋群岛),如果是真的,那么西方人把风筝尾巴做成鸢尾状就应该是从中国学去的。(刘夙)

马蔺

鸢尾科鸢尾属
Iris lactea (Chinese iris)
Iridaceae

又称马莲。多年生宿根草本，分布于我国西北、华北。单叶基生，条形，带红紫色，无明显的中脉。花期4–6月，花茎光滑；花较小，蓝紫色，外轮花被片3，弯曲下垂，内轮花被片3，小而直立；花柱花瓣状，顶端2裂。果期6–9月，蒴果，长圆柱形。未名湖北岸路边石头周围多有栽培，作观赏植物。北京市地名"马连道""马连洼"即来自此种植物的别名。

Perennial herb with strong rhizome and root systems. This species is native to North and Northwest China, Korea, and Central Asia. It can grow in very harsh habitats, and is often used as ground cover. It is sporadically found on campus, with one under the Bo Ya Pagoda.

黄菖蒲

鸢尾科鸢尾属
Iris pseudacorus (yellow iris)
Iridaceae

又称黄花鸢尾。多年生湿生宿根草本。单叶基生，宽条形，先端渐尖，有3–5条不明显的纵脉。花期5–6月，花茎中空，有1–2枚茎生叶；花黄色，外轮花被片3，具紫褐色的条纹，两侧边缘有紫褐色耳状突起物，内轮花被片3，倒披针形，花盛开时向外倾斜。果期7–8月，蒴果，椭圆状柱形。观赏植物，未名湖西边水渠边有栽培。

Perennial herb, native to central and Southwest China. This species can adapt to various habitats, and is often cultivated as ornamental. The roots are used as herbal medicine. It is cultivated at the western end of Wei Ming Lake.

德国鸢尾

鸢尾科鸢尾属
Iris germanica (German iris)
Iridaceae

多年生宿根草本，原产于欧洲，单叶基生，剑形，无明显的中脉，常具白粉。花期 5–6 月，花大而美丽，淡紫色、蓝紫色、深紫色或白色，有香味，外轮花被片 3，反折，具条纹，中脉上密生黄色须毛状附属物，内轮花被片 3，上部向内拱曲；花柱分支扁平，花瓣状。果期 7–8 月，蒴果，三棱状圆柱形。为世界各地最常见的栽培鸢尾花卉之一，有很多品种。未名湖东岸、红湖西岸有栽培。

Perennial herb, native to Europe. This is the most commonly found garden iris with numerous cultivars. The rhizomes are traded as orris root and are used in making perfume and medicine, though more common in ancient times than today. It is planted on east side of Wei Ming Lake, and around Hong Hu on the west side of campus.

独行菜

十字花科独行菜属
Lepidium apetalum (pepperweed)
Brassicaceae

一至二年生草本，分布于我国北方。根有特殊辣味。茎直立有分支，具头状短柔毛。单叶，基生叶羽状裂，有叶柄；茎生叶互生，基部宽，无柄，呈耳状抱茎，最上部叶线形。花果期4–6月，总状花序顶生；花两性，极小，无花瓣，雄蕊2或4。果实为短角果，近圆形。常见杂草，北京分布极多，校园草地、路边皆有分布。

Annual or biennial herb, native to China, India, Japan, Kazakhstan, Korea, Mongolia, Nepal, and Pakistan. This species is a common weed, usually found on roadsides, slopes, waste places, ravines, plains, and fields. The extract of the seeds could be used as a cardiac stimulant and antiasthmatic. It grows in large populations along roadsides on campus.

早园竹

禾本科刚竹属
Phyllostachys propinqua
(propinquity bamboo)
Poaceae

常绿乔木，原产于我国中部和东部。节间全为绿色，新秆被厚白粉，或有时仅节下有白粉环，秆环与箨环均中度隆起。笋期4—5月，秆箨无箨耳和肩毛。叶片披针形，宽2—3厘米。临湖轩有栽培。

This is an evergreen bamboo species native to central and East China. It is cultivated widely as an ornamental. The culms of this hardy species are used for weaving and for tool handles. A large population is found in front of Lin Hu Xuan.

北方的竹子在筋骨上不入流，但风色却有独到的地方，所谓"绿肥"。这在下雪天就格外精神，森郁的竹丛，冷碧的叶子上承着厚雪，很能激发文人之想。难怪当年在燕大的冰心选在这里举行婚礼，她的文字那么晶莹明爽，就像被雪澡过的竹叶。才高如张爱玲，也得暗服冰心的真。（王立刚）

燕园草木

乳浆大戟

大戟科大戟属
Euphorbia esula (green spurge or leafy spurge)
Euphorbiaceae

又称猫眼草。多年生直立草本，我国大部分省区有分布。雌雄同株，具白色乳汁，茎通常分支，基部坚硬。单叶互生，叶形变化大，一般为线状披针形，两面无毛。花期 4–6 月，杯状聚伞花序顶生或腋生，各有扇状半圆形或三角状心形苞叶 1 对。果期 6–8 月，蒴果，扁球形，无毛。常见杂草，校园有野生。

Perennial herb, native in Eurasia. It is naturalized in North America and classified as a noxious weed there. The plant is toxic, but after processing could be used as an expectorant. The seed oil is used in industry. It is a common weed on campus.

乌头叶蛇葡萄

葡萄科蛇葡萄属
Ampelopsis aconitifolia
(aconite-leaf ampelopsis)
Vitaceae

落叶木质藤本。幼枝被黄绒毛；卷须与叶对生。掌状复叶互生，轮廓为广卵形，小叶3-5，全部羽裂或深裂，边缘有大圆钝锯齿，有时小叶更细裂成条形（即变种掌裂草葡萄）。花期4-6月，聚伞花序与叶对生；花小，两性，黄绿色，5数，雄蕊与花瓣对生。果期7-10月，浆果，近球形，成熟时橙黄色。校园较多野生。未名湖南岸山坡路边很常见，常攀于树上或灌丛上。

Deciduous woody vine, native to central, North, and Northeast China. This species is sometimes planted along the edge of woods, or for shade on wooden corridors in gardens or parks. It is a common weedy vine on campus.

异穗薹草

莎草科薹草属
Carex heterostachya
(heterostachy sedge)
Cyperaceae

多年生草本，分布于我国北方。雌雄同株。茎三棱形。单叶，多为基生，线形，基部具闭合包茎的褐色叶鞘。花果期4-6月，小穗3-4个，顶生小穗为雄性，侧生的其他小穗为雌性。果实为小坚果，三棱形。本种可作草坪草或一般地被绿化植物，校园多有生长，如老生物楼前东西两侧林下和塞万提斯及蔡元培塑像前的草地上。

Perennial herb, native to North and Northeast China and Korea. This species could be used as lawn grass, and animal feed. It is a common species on campus.

臭草

禾本科臭草属
Melica scabrosa (rough melica)
Poaceae

多年生草本，分布于我国华北和西北。叶鞘闭合，下部叶鞘长于节间，上部叶鞘短于节间；叶舌膜质透明，顶端撕裂。花期4-7月，圆锥花序窄狭，分支紧贴主轴；小穗轴顶端几朵退化不育小花集合成球状。果期5-8月，果实为颖果。杂草，校园多见，凡草地或荒地、路边均有。

Perennial herb, native to North China, Korea, and Mongolia. This species is usually distributed on rocky slopes and graveled river banks. It is a common weed on campus.

夏至草

唇形科夏至草属
Lagopsis supina
(white-flower lagopsis)
Lamiaceae

多年生直立草本，我国分布广泛。茎四棱形，密被柔毛。单叶对生，叶片掌状3全裂，裂片有钝齿，两面均密生细毛。花期3-6月，轮伞花序；花两性，花萼钟形，具5齿，齿端有尖刺，花冠白色，二唇形，上唇较下唇长；雄蕊4,2长2短，不伸出。果期6-7月（夏至前后），结果时地上部分即枯萎，小坚果4。杂草，校园很多，如一教前绿篱下即有。

Perennial herb, native to central, North, and Northeast China, Japan, and Far East Russia. The whole plant is used medicinally. It is commonly seen on campus.

蛇莓

蔷薇科蛇莓属
Duchesnea indica
(mock strawberry or false strawberry)
Rosaceae

多年生匍匐草本，分布于我国北方。具长匍匐茎，被柔毛。羽状复叶互生，具3小叶，叶柄长；小叶菱状卵形，边缘具钝锯齿。花期4-7月，花两性，单生叶腋，除5枚萼片外，还具5枚大于萼片的副萼，边缘3浅裂，花后反折，花瓣5，黄色，与萼片等长；雄蕊多数，比花瓣短。果期5-10月，聚合果，暗红色，形似草莓而小。蔡元培像附近有分布。果实可食，但无甜味；可驯化作地被植物。

Perennial herb with robust short rhizomes. This species is distributed in most parts of China, and in many countries in Asia, and naturalized in Africa, Europe, and North America. It is considered a noxious weed in some regions, but it is used for ground cover in some other regions. The whole plant is used medicinally, mostly for reducing inflammation. It was recently introduced on campus, on the lawn near the statue of Cai Yuanpei.

刺儿菜

菊科蓟属
Cirsium setosum (setose thistle)
Asteraceae

多年生草本，除西藏外全国皆有分布。雌雄异株。茎直立，幼茎被白色蛛丝状毛。单叶互生，下部和中部叶椭圆形，两面有疏密不等的白色蛛丝状毛，近全缘或有疏锯齿，齿端有刺。花果期4-8月，头状花序直立，总苞片6层，有刺；花全为管状花，花冠紫红色。瘦果椭圆形，冠毛羽毛状。常见杂草，校园草地、荒地有野生。

Perennial herb, widely distributed in temperate zones of Eurasia. It is a common weed on wastelands, roadsides, edges of farmlands. The young plant could be used as a vegetable, and the whole plant is used medicinally. It is found along roadsides on campus.

平车前

车前科车前属
Plantago depressa
(depressed plantain)
Plantaginaceae

一年生草本，全国皆有分布。具明显的圆柱形直根，须根少。单叶基生，长圆状披针形，基出脉 3-7 条。花果期 4-8 月，穗状花序直立；花两性，小而不明显，萼片和花冠均 4 裂。果实为蒴果，圆锥状，盖裂。为极常见的伴人杂草，校园多见，凡草地、路边皆有。

Annual herb, distributed to China, Central Asia, and East Russia. This is a weedy species, and the whole plant is used medicinally. It is a common weed on campus.

圆叶鼠李

鼠李科鼠李属
Rhamnus globosa (lokao buckthorn)
Rhamnaceae

落叶灌木，分布于我国长江以北。枝灰褐色，小枝细长，枝端锐尖成刺，小枝、叶脉、叶柄均被毛。单叶互生或近对生，纸质，倒卵形或近圆形，边缘有钝锯齿，两面均有短毛，侧脉 3-5 对。花期 5-6 月，聚伞花序；花小，簇生于叶腋，雄蕊与花瓣对生。果期 8-9 月，核果，近球形。校园多见，尤以博雅塔附近未名湖南侧的山坡边为多。

Deciduous shrub, endemic to North, Northeast, and East China. Oil extracted from the seeds is used for making lubricating oil; and the bark, fruits, and roots are used for making a green dye. It is a very common shrub on campus, especially on the hill south of Wei Ming Lake.

灰栒子

蔷薇科栒子属
Cotoneaster acutifolius
(sharp-leaf cotoneaster)
Rosaceae

落叶灌木，我国分布广泛，北京有栽培。老枝灰黑色，嫩枝被长柔毛。单叶互生，小形，卵形，全缘，幼时两面有长柔毛。花期5-6月，聚伞花序；花小，萼筒外被柔毛，花瓣5，直立，近圆形，白色；雄蕊多数。果期8-9月，梨果，倒卵形或椭圆形，鲜红色至紫黑色，被疏毛，有2-3小核。观赏植物，办公楼礼堂东北角三角地上有栽培。

Deciduous shrub, native to China, Mongolia and Russia. This species is cultivated for its dense flowers in spring and red fruit in fall. It is found near the Administration Building on campus.

二球悬铃木

悬铃木科悬铃木属
Platanus acerifolia
(London plane or hybrid plane)
Platanaceae

又称英国梧桐。落叶乔木，可能是一球悬铃木（美国梧桐）和三球悬铃木（法国梧桐）的杂交种。我国广泛栽培。雌雄同株。树皮光滑，片状剥落。单叶互生，阔卵形，3-5裂，像枫叶，边缘疏生牙齿；托叶较长。花期5月，雄花和雌花均密集成头状花序，生于不同的花枝上。果期9-10月，果枝上的球形果序通常为2个，偶尔只有1个。为世界闻名的行道树种，校园有栽培，以文史楼南侧两株最大，而原浴室南侧及原校医院对面也有集中分布。

Deciduous tree. This plant is either a hybrid between *Platanus occidentalis* and *P. orientalis* or a cultivar of *P. orientalis*. It is widely cultivated in China as shade trees along the sidewalk of strees. Several big trees are found in front of the Chinese and History Building, the Administration Building, and the Campus Dining Complex.

浴室南面的这排美丽的乔木生长在北大最热闹的地段，多少女孩的雨伞上曾经落过它巨大的叶子，多少男孩的短发上曾经落过它滤下的雨滴。多少个酷夏，人们从它们脚下获得短暂的清凉，多少次冲澡，对它们"坦诚相见"。（王立刚）

毛梾

山茱萸科山茱萸属
Cornus walteri (Walter dogwood)
Cornaceae

又称车梁木。落叶乔木，我国南北各省有分布。树皮黑褐色，小枝绿白色。单叶对生，长椭圆形，全缘，两面都生有短柔毛，侧脉弧形，4–5对。花期5月，伞房状聚伞花序顶生；花两性，白色，4瓣。果期9月，核果，球形，黑色。俄文楼西侧草地西边有一株。

Deciduous tree, native to China. The leaves and fruit are a source of animal feed; the hard wood is used for making tools, and the tree itself is planted as a street tree. One tree is planted on the east side of Nan Ge.

红瑞木

山茱萸科山茱萸属
Cornus alba (red-barked dogwood)
Cornaceae

落叶灌木，分布于我国北方。树皮紫红色，幼枝有淡白色短柔毛。单叶对生，纸质，椭圆形，全缘，下面被白色贴生短柔毛，侧脉弧形，4-6对，两面均明显。花期6-7月，伞房状聚伞花序顶生；花小，两性，淡黄白色，4数。果期8-10月，核果，长圆形，成熟时乳白色或蓝白色，有宿存花柱。常见观赏植物，勺园南边及北部假山下草地有栽培。

Deciduous shrub, with purplish red bark. It is native to North and Northeast China, Korea, Mongolia, Russia, and Europe. This is a popular ornamental shrub used in landscaping. The seeds contain 30% oil, which is used industrially. Recently it was cultivated near Shao Yuan.

毛洋槐

豆科刺槐属
Robinia hispida
(bristly locust, or rose acacia)
Fabaceae

落叶灌木，原产于北美，北京、上海等城市有栽培。小枝、花梗及叶柄密被棕褐色刚毛。奇数羽状复叶互生，小叶 7–15，近圆形或长圆形，全缘。花期 5 月，总状花序腋生；花两性，蝶形，粉红或紫红色；雄蕊 10，二体。一般不见结果。观赏植物，红二、三楼的南边草地上栽有多株。

Deciduous shrub, native to Central and Eastern America. It is cultivated in China mostly for its showy purplish flowers. It is planted in front of Cai Zhai.

火棘

蔷薇科火棘属
Pyracantha fortuneana
(fortune firethorn)
Rosaceae

常绿灌木，原产于我国华中、华东及西南各省，广为栽培。侧枝短，顶端刺状。叶互生，倒卵形，先端圆钝或微凹，边缘有圆钝齿，叶基楔形，下延；叶柄短；花期5月，两性花集成复伞房花序，花白色，萼筒钟状，无毛，花瓣圆形。果期8-9月，梨果近球形，富含维生素，可做饮料的原料。校史馆北侧栽培数株，深秋及初冬时橘红色的果实非常引人注目。

Deciduous shrub, native to China. This species is often cultivated in gardens and parks for its springtime flowers and bright orange or red fruit in the fall and early winter. The fruit, rich in vitamins, is used as raw material for soft drinks. It is planted in front of the Museum of Peking University History.

欧洲荚蒾

五福花科荚蒾属
Viburnum opulus subsp. *opulus* (European cranberry bush or snowball tree)
Adoxaceae

为欧洲荚蒾的原亚种，落叶灌木，原分布于欧洲及我国浙江省的西北部山上。树皮薄；单叶对生，阔卵圆形，常3浅裂，叶上面皱，具掌状3出脉，1至4级脉常处于凹陷处；花期5-6月，伞形聚伞花序顶生；花有二形，原亚种花序外围仅有一圈不育花，白色，花冠深5裂，中间为小形的可育花，花蕾绿白色，花药黄白色。果期8-9月，核果，球形，鲜红色。其品种"蔷薇"欧洲荚蒾的花序球状，均为不育花，北京市区常见栽培。观赏植物，校园西部有栽培，北阁北边林下及勺园5号楼后栽培有原亚种，其他地点零星栽培有"蔷薇"欧洲荚蒾。

Deciduous shrub. This subspecies of *Viburnum opulus* is native to Europe, and Northwest of Zhejiang Province, China. The outer flowers in the inflorescence are sterile. The leaves, young branches, and fruit are used medicinally. Some cultivars have sterile flowers in the whole inflorescence. This subspecies and some of its cultivars, such as "Roseum", are planted on campus.

互叶醉鱼草

马钱科醉鱼草属
Buddleja alternifolia
(fountain butterfly bush)
Loganiaceae

落叶灌木，分布于我国西北各省。枝细弱，多呈弧形弯曲。单叶互生，披针形，全缘，下面密被灰白色柔毛及星状毛。花期5-6月，圆锥花序，簇生于二年生枝叶腋；花冠筒状，紫蓝色或紫红色，裂片4；雄蕊4，无花丝。果期6-7月，蒴果，长圆状柱形。静园一院南边花园中栽培。

Deciduous shrub, native to Northwest China. The flowers are rich in essential oil. Only one plant is found on campus, southeast of Jing Yuan.

杠柳

萝藦科杠柳属
Periploca sepium
(Chinese silkvine)
Asclepiadaceae

落叶木质藤本，广泛分布于我国南北各省。除花外全株无毛，具白色乳汁。单叶对生，革质，全缘，侧脉多数。花期6-7月，二歧聚伞花序腋生；花紫红色，具环状副花冠，被柔毛。果期8-9月，蓇葖果双生；种子多数，顶端具种毛。8公寓水域对面山坡上有野生，朗润园一带山上也有。

Deciduous woody vine, native to China. This species has a strong root system making the plant tolerant to drought conditions. It is often planted on slopes for preventing erosion. Its bark and root could be used as medicine. It grows sporadically on campus.

二体篮球场整修之前，护栏上曾经爬满了杠柳。细长柔韧的茎干在护栏的网格上缭乱地书写，一两枝忽地向着头顶那一片湛蓝去了，舒展的叶片流光溢彩，间或缀几朵张牙舞爪的小花，奶黄或是粉紫。篮球声和加油声就从叶片间泄露出来。而如今球声依旧，却不见那昔日的绿荫。（蔡乐）

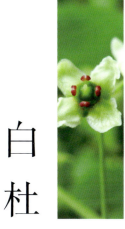

白杜

卫矛科卫矛属
Euonymus maackii (winterberry euonymus or Maack spindle-tree)
Celastraceae

落叶灌木或小乔木，我国北方广泛分布。小枝灰绿色，圆柱形。单叶对生，卵状椭圆形，边缘具细锯齿。花期5-6月，聚伞花序3至多花；花两性，4瓣，淡黄绿色；雄蕊花药紫红色，花丝细长。果期9月，蒴果，倒圆心状，4浅裂，成熟后果皮粉红色；种子外被橘红色假种皮。未名湖边南岸、东岸均有较老的树。

Deciduous tree. It is native to the northern and middle parts of China, in Japan, Korea, Russia (Far East), and also cultivated in Europe and North America. The roots, barks, leaves, and fruit are used as Chinese traditional medicine. It is often cultivated as a shade tree along sidewalks or ornamental in gardens. Several trees are planted along Wei Ming Lake.

红丁香

木犀科丁香属
Syringa villosa (late lilac)
Oleaceae

落叶灌木，分布于我国华北和东北。枝直立，具皮孔。单叶对生，叶片椭圆状卵形，常贴生疏柔毛，背面被白粉，全缘。花期5–6月，圆锥花序顶生；花两性，芳香，花冠淡紫红色、粉红色至白色，高脚杯状，花冠管细弱；雄蕊2。果期9月，蒴果，长圆形，2裂。野生于较高海拔地区，未名湖南岸新近引种栽培。

Deciduous shrub. This species is endemic to North China. It usually grows under shade, and is often cultivated as ornamentals. It has recently been planted along the southern part of Wei Ming Lake.

雪柳

木犀科雪柳属
Fontanesia philliraeoides subsp. *fortunei*
Oleaceae

落叶灌木，原产于我国北方及东部。小枝四棱形，无毛。单叶对生，叶片纸质，披针形，似柳叶，但全缘，两面无毛。花期5-6月，圆锥花序，顶生或腋生；花两性或杂性同株，白绿色，有香味。果期8-9月，翅果，卵圆形，扁平。观赏植物，电教东边有一排老树，生物技术楼前种有数棵。

Deciduous shrub or small tree. This species has two subspecies. Subspecies *fortunei* is endemic to North and East China, and cultivated as an ornamental. It is found west of the Second Teaching Complex, and in front of the Biotechnology Building. Another subspecies *philliraeoides* is native to Southwest Asia and Europe.

红花锦鸡儿

豆科锦鸡儿属
Caragana rosea
(red-flower pea shrub)
Fabaceae

直立灌木，分布于我国长江以北各省，北京山区分布普遍。小枝细长，具刺。小叶4，假掌状排列，椭圆状倒卵形。花期5–6月，花两性，单生，花萼筒状，萼齿三角形，有刺尖，花冠蝶形，黄色，常带紫红或淡红色，凋时变为红色；雄蕊10，二体。果期6–8月，荚果，圆筒形，具尖刺。档案馆东边山林内近路边有一大丛。

Deciduous shrub. This species is endemic to North and East China. The roots are used for herbal medicine as a diuretic and an antiasthmatic. Only one plant is found east of the university Archive Building.

锦带花

落叶灌木，分布于我国北方。幼枝有柔毛。单叶对生，椭圆形，边缘有锯齿，两面被短柔毛。花期5-6月，伞房花序在侧枝上顶生；花两性，花萼5裂，被毛，花冠漏斗状钟形，外面粉红色，里面灰白色；雄蕊5。果期10月，蒴果，柱状。观赏植物，有很多栽培品种，校园东门附近和蔡元培塑像周围种有"花叶"锦带花，生物技术楼附近种有"红王子"锦带花。

Deciduous shrub, native to North and Northeast China. This species is widely cultivated in gardens and parks with many different colored flower cultivars. The dense flowers can last for several months. Several cultivars, such as *Weigela florida* "Varieagata" with white-green leaves and "Red Prince" with brightly red flowers, are planted near the East Gate, statue of Cai Yuanpei, and the Biotechnology Building.

忍冬科锦带花属
Weigela florida
(old-fashioned weigela)
Caprifoliaceae

金银忍冬

又称金银木。落叶灌木，我国分布广泛。小枝中空。单叶对生，卵状椭圆形，全缘，两面疏生柔毛。花期 5-6 月，花成对腋生，两性，花冠二唇形，初开时白色，后变黄色；雄蕊 5。果期 8-10 月，浆果，球形，半透明、亮红色。常见观赏植物，冬季红果满树，十分美丽，校园多有栽培。

Deciduous shrub, native to temperate East Asia. This species was naturalized in New Zealand and the United States, and has become an invasive species there. It is widely cultivated for its white and yellow flowers, and bright red and semi-translucent berries. It is very common on campus.

忍冬科忍冬属
Lonicera maackii
(bush honeysuckle)
Caprifoliaceae

君迁子

柿树科柿树属
Diospyros lotus (date-plum or Caucasian persimmon)
Ebenaceae

又称黑枣。落叶乔木，我国北方多见。雌雄异株。树皮黑灰色，裂成方块状。单叶互生，椭圆形，全缘。花期5-6月，花单生，淡黄色至淡红色，花萼和花冠裂片均为4。果期10-11月，浆果，近球形至椭圆形，明显小于柿子，初熟时淡黄色，后变为蓝黑色，具宿存花萼。可以作为嫁接柿树的砧木，红一楼后停车场有棵雄树，生物技术楼西北湖边有几株雌树。

Deciduous tree, native to subtropical Southwest Asia and South Europe. This species is naturalized in countries around the Mediterranean, and cultivated in the United States and North Africa. The fruit is edible and rich in sugars and vitamins. One male tree is found in the parking lot by the Arthur M. Sackler Museum of Art and Archaeology, and several female trees are at the northwestern end of campus.

柿

柿树科柿树属
Diospyros kaki
(Asian persimmon, or kaki persimmon)
Ebenaceae

落叶乔木，全国均有分布，但以华北最多。雌雄异株。树皮黑灰色，裂成方块状。单叶互生，椭圆形或倒卵形，全缘，背面有绒毛。花期5-6月，花单生，黄白色，花萼和花冠裂片均为4。果期9-10月，浆果，橙黄色或黄色，具宿存花萼。为著名果树，果实可鲜食，也可制作柿饼。未名湖北岸及学生宿舍区有栽培，据说果不好吃，但叶子变成红色时非常好看。

Deciduous tree, endemic to China. This species is widely cultivated both in China and outside China, especially in Japan with many cultivars, and naturalized in some areas. It is among the oldest plants in cultivation, known for its use in China for more than 2000 years. Its fruit is famous fruit and rich in vitamins and soluble sugars. Its reddish foliage and orange-colored fruit are also an attraction in the fall. It is cultivated north of Wei Ming Lake and around the student dormitories.

关于燕园最美丽而遗憾的记忆应该有三角地的柿子林。每当霜白而秋实，万柿如灯，说不出的璀璨和温暖。它散发的是蓬勃之气。高挂的柿子总让我想起五六十年代的宣传画里青年们的脸颊。那种气色是如此饱满，就像是神在他们的灵魂里涂了一层金子。（王立刚）

毛泡桐

玄参科泡桐属
Paulownia tomentosa
(empress tree or foxglove tree)
Scrophulariaceae

落叶乔木，北京常见。树皮褐灰色，小枝有时因患丛枝病而极密集丛生。单叶对生，大形，卵状心形，全缘或具 3-5 浅裂，叶柄常有黏性腺毛。花期 5-6 月，聚伞圆锥花序；花萼浅钟状，密被星状绒毛，5 裂至中部，花冠漏斗状钟形，外面淡紫色，有毛，内面白色，有紫色条纹。果期 8-9 月，蒴果，卵圆形，萼片宿存。老生物楼东北方向路边有数株。

Deciduous tree, native to central and West China. This is a fast growing species and cultivated world-wide, and considered as an invasive species in the United States. Carving the wood of this tree is an art form in Japan and China. It is planted near the School of Journalism and Communication on compass.

三裂绣线菊

蔷薇科绣线菊属
Spiraea trilobata
(three-lobed spiraea)
Rosaceae

落叶灌木，分布于我国长江以北。小枝开展，呈之字形弯曲，无毛。单叶互生，近圆形，先端3裂，边缘自中部以上具少数圆钝锯齿，两面无毛。花期5-6月，伞形花序；花小，两性，花瓣5，白色，先端微凹；雄蕊多数，比花瓣短。果期7-8月，果实为蓇葖果。为华北地区极常见的阳坡灌丛植物，蔡元培塑像草地西南角有栽培。

Deciduous shrub, native to China, Korea, and Russia. The leaf shape of this species is very variable. It is often cultivated as an ornamental. It is planted near the statue of Cai Yuanpei.

山楂

蔷薇科山楂属
Crataegus pinnatifida
(Chinese hawthorn)
Rosaceae

落叶灌木或小乔木,著名果树,分布于我国长江以北。枝密生,有细刺,幼枝有柔毛。单叶互生,三角状卵形至菱状卵形,两侧各有3-5枚羽状深裂片,边缘具不规则锐锯齿。花期5-6月,复伞房花序,花序梗、花柄都有长柔毛;花两性,花瓣5,白色;雄蕊20。果期9-10月,梨果,深红色,近球形,可食用。校史馆东侧林中有棵大树,哲学系院内栽有其变种山里红(*C. pinnatifida* var. *major*)。

Deciduous tree, native to North and East China, Korea. One of the varieties of this species, *C. pinnatifida* var. *major*, is widely cultivated in northern China as a fruit tree. The matured red fruit is the material for the famous local dessert "tang-hu-lu", and it is used to make the traditional haw flakes, as well as candied fruit slices, jam, jelly and wine. It is also used as a basic material for some traditional Chinese medicine for helping digestion. A tree of this species is found in the woods east of the university Archive Building, and var. *major* is cultivated in the yard of the Department of Philosophy.

北大五院有株山楂颇特别……同一株树上竟然不同时开花，结出的果子也不一样！……一种是稀疏、个头很大的普通大山楂，另一种很像在东北山上常见的山里红。可以合理地猜测：当初苗圃园工可能用一株同科同属不同种的野生山楂的小苗作砧木，嫁接成活了一株大山楂树苗……它顺利地在北大五院成长，并结出美丽、可食的果子。（刘华杰）

流苏树

木犀科流苏树属
Chionanthus retusus
(Chinese fringe tree)
Oleaceae

落叶灌木，分布于我国南北各省。雌雄异株。枝开展，小枝灰绿色。单叶对生，椭圆形，全缘，背面常具柔毛。花期6-7月，聚伞状圆锥花序着生侧枝枝顶；花冠白色，深4裂，裂片线状倒披针形；雄花具2雄蕊。果期9-10月，核果，椭圆形。观赏植物，鸣鹤园湖边、塞万提斯塑像周围种有多棵。

Deciduous shrub or small tree, native to China, Japan and Korea. This species is cultivated in Europe and North America as an ornamental tree, valued for its feathery white flowers. It is planted around Ming He Yuan and the statue of Cervantes.

在承泽园里有一株流苏树，属木犀科，是珍稀之种。此树有2干，高15米，为老树，春天开花时为一树白雪，其花冠细裂如条，形态极为奇特，何年所栽已难考证，反正是新中国以前的遗物。（汪劲武）

大花糯米条

北极花科糯米条属
Abelia × grandiflora
(big-flower abelia)
Linnaeaceae

半常绿灌木，是蓪梗花（*A. uniflora*）与糯米条（*A. chinensis*）的杂交种，在我国和欧洲、美洲和非洲广为栽培。枝被短柔毛；叶对生，具稀疏缺刻状锯齿。花期5-7月，花成簇着生于侧枝末端，单个花着生在4个叶状的花萼裂片中，花瓣白色，有时略带粉红色。果期7-8月，瘦果。老生物楼东面山坡上有一株。

Semi-evergreen shrub. This hybrid between *Abelia uniflora* and *A. chinensis* is more commonly cultivated in the America, Africa, and Europe than in China. A small shrub was recently planted east of the former Biology Building.

落叶乔木。雌雄异株。全株具刺，刺常分支，且多密集。叶互生，1-2回羽状复叶，一回羽状复叶常簇生，二回羽状复叶具2-6对羽片，小叶3-10对，卵形或卵状长圆形。花期5-6月，雄花成细长的总状花序，雌花则为穗状花序。果期6-10月，荚果，带状，常不规则扭转。红二楼与红三楼之间的北侧有一株雌性老树，档案馆东侧山林中有一雄株，均年年开花。

Deciduous tree, native central, North, and East China, Japan, and Korea. This species is commonly cultivated in China. The heartwood has a beautiful pink color, and is used for furniture and building. The seed is used medicinally, and its boiled soup as detergent. One female tree is found between Cai Zhai and Jun Zhai, and one male tree is in the woods east of the university Archive Building.

山皂荚

豆科皂荚属
Gleditsia japonica
(Japanese honey locust)
Fabaceae

蝟实

忍冬科蝟实属
Kolkwitzia amabilis (beauty bush)
Caprifoliaceae

落叶灌木，中国特有植物，现世界各地有栽培。单叶对生，椭圆形，近全缘至疏具浅齿，两面具毛。花期6月，聚伞花序具2花，2花的萼筒下部合生；萼筒有长柔毛，花冠钟状，粉红色至紫色。果期7–9月，核果瘦果状，2个合生，有时其中1个不发育，外有刺刚毛，形似刺猬。优美的观赏植物，静园东北角栽培。

Deciduous shrub. This species is endemic to central China. Although it is rare in its native habitat, it is cultivated in gardens and parks world-wide for its dense flowers blooming in summer, and gracefully arching branches. It is cultivated in the northeastern corner of Jing Yuan.

猬实是一种优美的园艺植物。它的拉丁文种加词 amabilis 意为"可爱的",美国人据此也管它叫 beauty bush,1901年被引种到欧洲后,在庭园中广为栽培。比如在法国巴黎植物园的温室对面就有两株,每年五月底的时候,一树繁花,如灿烂的云霞。如果有细心的华人朋友上前查看标识牌,发现它竟然是来自 Hupeh(湖北),如今在万里之外的异域寂寞地开着美丽的花,不知会不会勾起心底的一丝乡愁?(刘夙)

杂种鹅掌楸

木兰科鹅掌楸属
Liriodendron chinense × *L. tulipifera*
(hybrid liriodendron)
Magnoliaceae

落叶乔木，为鹅掌楸属内仅有两个物种（中国鹅掌楸和北美鹅掌楸）的杂交种，栽培广泛。单叶互生，鹅掌形（或称马褂形），两侧各有1-3浅裂，先端近截形；托叶早落而留下环状托叶痕。花期5-6月，花两性，单生，浅黄绿色，花被外面绿色，内面黄色。通常不结果。该杂种具有明显的杂交优势，树形优美，健壮，在中国很多城市都有栽培。新生物楼周围近年来引种的杂种鹅掌楸长势很好，每年春天开花。

Hybrid liriodendron. This is the hybrid between *L. chinense*, a species native to East Asia, and *L. tulipifera*, a species native to North America. The trees display strong hybrid vigor, and are widely cultivated in Beijing. The Chinese name for this genus, goose-web tree, comes after its characteristic leaf shape. The tree was introduced to campus recently around the new Life Science Building.

卫矛

卫矛科卫矛属
Euonymus alatus (winged spindle or winged euonymus)
Celastraceae

落叶灌木，我国南北各省皆有分布。小枝通常四棱形，棱上常具木栓质扁条状翅。单叶对生，椭圆形，边缘有细锯齿。花期5月，聚伞花序腋生，有花3—9朵；花小，两性，4数，淡黄绿色；花丝短，着生于肥厚方形的花盘上。果期9—10月，蒴果，椭圆形，分离；种子外被橘红色假种皮。民主楼西侧智圣者塑像前和先进技术学院平房前皆有栽种。

Deciduous shrub, native to China, Japan, Korea, and Russia (Sakhalin). This species is cultivated worldwide. In the fall the leaves change from pinkish to reddish or purplish, and the seeds with bright red aril are exposed at maturation. The winged-branches are used medicinally. It is cultivated in Ming He Yuan and by the building of Advanced Technology Institute.

太平花

虎耳草科山梅花属
Philadelphus pekinensis
(Beijing mockorange)
Saxifragaceae

落叶灌木，我国分布广泛。幼枝光滑。单叶对生，卵形，边缘疏生锯齿，具3主脉，上面无毛。花期5-6月，总状花序，花两性，萼筒无毛，花瓣4，白色。果期8-9月，蒴果，倒圆锥形，4瓣裂。北大图书馆南门东侧草地的银杏树下多植，为观赏植物；北侧草地也有。

Deciduous shrub, native to North and Northeast China, and Korea. It has a long history of cultivation as an ornamental in gardens for its beautiful flowers and its name – "peace flower" in Chinese. Several plants are cultivated north of the library and northeast of the Administration Building under the gingko tree.

荇菜

睡菜科荇菜属
Nymphoides peltata
(floating bogbean)
Menyanthaceae

多年生水生草本，广布于我国南北各省区。茎圆柱形，沉水中，具不定根。单叶圆形，飘浮水面，近革质，基部心形，上部的叶对生，其他的为互生；叶柄抱茎。花果期6—8月，花两性，成束生于叶腋，黄色，花萼5深裂，花冠5深裂，喉部具毛；雄蕊5。果实为蒴果，长椭圆形。未名湖水域中引进的荇菜近年来长势非常好。

Perennial aquatic plant with a long creeping rootstock. This species is distributed in China, other parts of Asia, and Europe. It is commonly planted in ponds and lakes as an ornamental. The leaves are edible, and also used as animal feed. In recent years the introduced plants have grown into a large population on Wei Ming Lake.

《诗经·关雎》写到水中荇菜漂浮在美女左右，自然而然地衬托着佳人的优美身段。这图景令青年浮想联翩，夜不能寐。古时候荇菜的地位想必相当于今日的玫瑰。未名湖和朗润湖就有荇菜，静静地飘浮在湖边水面上，每年6月都如期开出漂亮的金黄色小花。……夏日里，校园的恋人们坐在湖边……我担保，荇菜的叶和花绝对值得仔细观赏。恋爱时想想《关雎》，也并不跑题。（刘华杰）

小花扁担杆

椴树科扁担杆属
Grewia biloba var. *parviflora*
(small flower grewia)
Tiliaceae

又称孩儿拳头。落叶灌木，我国分布广泛。小枝有星状毛。单叶互生，狭菱状卵形或狭菱形，边缘密生小牙齿，基出3脉。花期6-7月，聚伞花序与叶对生；花淡黄绿色，花瓣5；雄蕊多数。果期8-9月，核果，橙红色，2裂，形似小孩拳头。俄文楼西北的林中有许多。

Deciduous shrub or small tree, native to China and Korea. This variety is drought-tolerant, and often cultivated for horticulture. Its bark is a good resource for fiber, and its roots and leaves are used medicinally. The fruit turns into bright red when mature. It is found around Wei Ming Lake, and in the Ming He Yuan areas.

燕园草木

月季花

蔷薇科蔷薇属
Rosa chinensis (Chinese rose)
Rosaceae

常绿或落叶灌木，原产于我国，现在世界各地广泛栽培，品种甚多。小枝具有钩状皮刺。奇数羽状复叶，小叶3-5，两面无毛，具尖齿。花期4-10月，花单生，或数朵聚生成伞房状，花色很多，色泽各异，多为重瓣也有单瓣者；萼片卵形，先端尾尖，羽状分裂；花瓣倒卵形，先端外卷；花柱分离，长为雄蕊之半。果期9月，蔷薇果，卵形。月季是北京市市花，校园栽培有很多品种。

Evergreen or deciduous shrubs, native to central China. The species is extensively cultivated in China as an ornamental, and numerous cultivars with double flowers of various colors have been selected. The flowers and fruit are used in traditional Chinese medicine. Many cultivars are seen on campus.

玫瑰

蔷薇科蔷薇属
Rosa rugosa (rugose rose)
Rosaceae

直立灌木。茎丛生，有皮刺；奇数羽状复叶互生，小叶 5-9 片，椭圆形，先端急尖或圆钝，上面多皱，下面灰绿色，边缘有尖锐锯齿；花期 5-7 月，花单生枝顶或 2-6 朵聚生，极香，萼片 5，多扩大为叶状，花瓣 5，或重瓣，花柱离生；果期 8-9 月，蔷薇果扁球形，光滑无毛，萼片宿存。博雅塔下山坡路边有栽培。

Deciduous shrub, native to China, Japan, and Korea. This species is widely cultivated world-wide with many forms of cultivars used as ornamentals and sources of essential oil. The wild plants are ranked as "endangered" in *China Plant Red Data Book*. Several cultivars are planted on campus.

如果你没有收到过玫瑰（指月季），恐怕也送出过。也就是说，我们对玫瑰并不陌生，但玫瑰却有一个秘密，它的花萼通常有 5 个萼片，其中两片的两侧光滑，没有附属物；另两片两侧均有"小翅膀"；剩下的一片，它长得跟谁都不一样，一侧有翅另一侧无翅！（刘华杰）

月季、玫瑰的区别

月季花色多种，蔷薇果卵圆形或梨形，小叶多为3—5，少7，小叶上面平整；玫瑰花常为紫红色，极少白色，小叶5—9，上面多皱纹。

女贞

木犀科女贞属
Ligustrum lucidum (Chinese privet or broad-leaf privet)
Oleaceae

常绿乔木，分布于我国长江流域及以南各省区和甘肃南部，其他地区多有栽培；枝条有皮孔，全株无毛；单叶对生，革质而脆，椭圆形，全缘；花期 5–7 月，圆锥花序，花近无梗，花冠筒和花萼近等长，雄蕊和花冠裂片近等长；果期 7 月至翌年 5 月，核果矩圆形，紫蓝色。北京气候偏冷，冬天易被冻伤，但校园小南门女生宿舍旁种的一排树长势非常好。

Evergreen shrub or tree, endemic to China. This species is often cultivated as ornamental, and has been introduced to Europe and North America. The dried fruit has been used as a traditional medicine for a long time in China. Several trees grow very well in front of the student dormitory near the Southwest Gate.

暴马丁香

木犀科丁香属
Syringa reticulata (Asian tree lilac)
Oleaceae

落叶小乔木，分布于我国北方。枝条带紫色，有光泽。单叶对生，叶片卵形，基部通常圆形，两面无毛，全缘。花期6月，圆锥花序大而稀疏，常侧生；花白色，较小，有浓郁气味，花萼、花冠均4裂；雄蕊2。果期8-9月，蒴果长圆形，先端钝；种子周围有翅。著名观赏植物，在我国西北高寒地区的佛寺中多有栽培，作为菩提树的替代，故又名"西海菩提"。该种有两个亚种分布在中国：北京丁香和暴马丁香，在博雅塔下和原电话室门前有栽培。

Deciduous shrub or tree. This species is found in China, Japan, Korea and East Russia. Two of the three subspecies are found in China: *S. reticulata* subsp. *pekinensis* is endemic to central and North China, and subsp. *amurensis* is native to Northeast China, Korea and East Russia. This species is regarded as a holy tree in some Buddhist temples in Qinghai Province, China. The bark, trunks, and branches are used as an antiphlogistic drug and diuretic. The flowers are used in the preparation of perfumes. It is cultivated on the east side of campus, with two trees under the Bo Ya Pagoda.

萱草

百合科萱草属
Hemerocallis fulva (orange daylily)
Liliaceae

多年生宿根草本，原产于我国南部。单叶，基生，宽条形、对排成两列，背面有龙骨突起，嫩绿色。花期6月上旬至7月上旬，花呈顶生聚伞花序，可食用。花大而美丽，漏斗形，花被下部合成花被筒，上部具6裂片，长圆形，开展而反卷，边缘波状，橘红色。观赏植物，有多个变种和栽培品种，近年在未名湖边及电教东侧种植有园艺杂交品种"金娃娃"。

Perennial herb, native to China, India, Japan, Korea and Russia. This species is very variable with several cultivars and widely cultivated in China for its showy and edible flowers. It is also used medicinally. One of the cultivars, "Stella d'Oro", was recently planted around Wei Ming Lake and west of the Second Teaching Complex.

Hemero 是"白天"的意思，callis 是"美丽"的意思。二者合起来，是说它的花很美丽，但是只能开一个白天，早晨开放，到傍晚就枯萎了。中国古人认为食之可以忘忧，就有了"忘忧草"的美名。萱草还是母爱、母亲的象征，这也许是因为慈祥的母爱也可以起到使人忘忧的效果吧。（刘夙）

酢浆草

酢浆草科酢浆草属
Oxalis corniculata
(creeping lady's sorrel)
Oxalidaceae

多年生草本，我国广泛分布。全体有疏柔毛，汁液带酸味；茎匍匐或斜升，多分支。三出掌状复叶互生，小叶无柄，倒心形，有夜间闭合现象。花期5-9月，花两性，花瓣5，黄色。果期6-10月，蒴果，圆柱形，熟时种子自动从果实中弹出。常见杂草，办公楼礼堂西侧、南侧的石缝、墙缝中有生长。

Annual or short-lived perennial. This is a highly successful weedy and cosmopolitan species, particularly in areas disturbed by humans. Its leaves are edible with a taste of lemons. The entire plant is rich in Vitamin C and used medicinally. It is a common weed on campus.

草地早熟禾

禾本科早熟禾属
Poa pratensis
(smooth meadow grass)
Poaceae

多年生草本，分布于我国长江以北，但栽培品种多系从国外引进。具长而明显的匍匐根须茎。叶舌膜质，截形；叶片条形，光滑，扁平，内卷。花期5-6月，圆锥花序卵圆形，开展；小穗卵圆形，草绿色，成熟后草黄色，基盘有稠密的白色绵毛。果实为颖果。为北方主要的草坪草，校园栽培多。

Perennial herb, native to Eurasia and North Africa. This species grows well under moderately moist to wet conditions, often in disturbed sites in temperate zones; and is cultivated world-wide now. It is a valuable pasture plant and its cultivated forms are often used for making lawns in parks and gardens. Several patches are found north of the PKU Library, and along roadsides on campus.

风花菜

十字花科蔊菜属
Rorippa globosa (globate Rorippa)
Brassicaceae

一年生草本，我国南北有分布。茎直立，有分支，无毛。单叶互生，长圆形或倒卵状披针形，基部抱茎，两侧尖耳状，边缘齿裂，无毛。花果期6-8月，总状花序顶生；花两性，很小，黄色，花瓣4，成十字形；四强雄蕊。短角果球形，无毛，先端有短喙。常见杂草，校园西门水边和未名湖南岸水边都有分布。

Annual or short-lived perennial herb, native to East Asia. The whole plant is used as herbal medicine. It usually grows in moist places near lakes or ponds on campus.

一年或二年生草本，我国各地有分布。第一年生出基生叶，组成大形莲座状，叶片倒披针形，提琴状羽状分裂，顶裂片较大，三角形，侧裂片 7-8 对，叶背被白色蛛丝状毛。第二年生出茎，光滑或有白色蛛丝状毛。花果期 5-8 月，头状花序多数，总苞球形，背面顶端有紫红色鸡冠状附片；花全部为管状花，花冠紫红色。瘦果圆柱形，冠毛白色，羽毛状。杂草，校园多见，习生草地路边。

Annual or biennial herb. This species is widely distributed in China, Japan, Korea, South Asia, and Australia. The young leaves are used as animal feed, and the whole plant is used medicinally. It is commonly seen on campus.

泥胡菜

菊科泥胡菜属
Hemistepta lyrata (lyrate hemistepta)
Asteraceae

半夏

天南星科半夏属
Pinellia ternata (crow-dipper)
Araceae

多年生小草本，分布于我国长江以北地区。块茎近球形。叶出自块茎顶端，一年生叶为单叶，卵状心形；2-3 年后，叶为 3 小叶的复叶，全缘，两面光滑无毛。肉穗花序顶生，佛焰苞绿色，花单性，无花被，雌雄同株。雄花着生在花序上部，白色，雌花着生于雄花的下部，绿色。浆果卵状椭圆形，绿色。花期 5-7 月。果期 8-9 月，浆果，卵状椭圆形，绿色。静园一院北侧墙下草地、南阁附近草地、陈守仁国际研究中心西侧草地中发现过。

Perennial herb. This species is widely distributed in China, except in Northwest China. It is also found in Japan and Korea. The whole plant is toxic in raw form. However, processed tubers are used in traditional Chinese medicine for treating coughs, reducing phlegm, stopping vomiting. The plant is also used ornamentally, and grows as an invasive weed in parts of North America. It is sporadically found on campus.

半夏有毒，因一般在夏季采收，故名"半夏"。在北京多生于山区，却见于静园草坪，大概是因为培育草皮的苗床先种过半夏，一些块茎随着草皮来到了燕园。这次偶然的逸生让半夏远离山谷的苦寒，享受没有乔木遮挡的阳光,该不会有埋于荒秽之叹吧。（刘华杰、刘凤）

萹蓄

蓼科蓼属
Polygonum aviculare
(common knotgrass)
Polygonaceae

一年生草本，全国各地均产。高 10-40cm，常有白粉；茎丛生，绿色，有沟纹；叶互生，条形至披针形，全缘，基部楔形，近无柄；托叶鞘膜质，2 裂，下部带绿色，上部白色透明，有明显脉纹。花期 5-8 月，花 1-5 朵簇生叶腋，露出托叶鞘外；花被 5 裂；雄蕊 8；花柱 3 裂。果期 9-10 月，瘦果卵形，表面有棱。校园路边常见有野生。

Annual herb near fields, roadsides, and waste places. This species is widely distributed in the northern temperate zone, and naturalized in the southern temperate zone. It is a well-known traditional herbal medicine, and has long been used as an astringent, diuretic, expectorant, and pesticide in China. It is often found along roadsides on campus.

茴茴蒜

毛茛科毛茛属
Ranunculus chinensis
(Chinese bettercup)
Ranunculaceae

多年生湿生草本，我国南北各省均有分布。茎和叶柄密被伸展的带淡黄色长硬毛。叶基生或在茎上互生，下部叶为三出复叶，上部叶为3全裂。花期5-8月，单歧聚伞花序；花两性，萼片5，淡绿色，花瓣5，黄色。果期6-9月，聚合瘦果，椭圆形，长约1厘米。西校门南侧水池的南岸和陈守仁研究中心南草地有分布。

Perennial or annual herb, native to Asia, usually by streams, rivers and wet grassy places. The whole plant can be used medicinally for reducing inflammation and swelling. It is found near the lakes on campus.

马齿苋

一年生匍匐草本，我国长江以北分布广泛。植物体肉质，茎分支，平卧地面。单叶互生，偶尔对生，全缘，肉质。花期5–8月，花两性，黄色，在枝端顶生，花瓣5；雄蕊8–12。果期7–9月，果实为盖裂蒴果；种子多数，肾状卵圆形。校园荒地多有，为常见杂草。

Annual herb, distributed in disturbed urban sites throughout China. The common weed of cultivation is edible as a vegetable and used medicinally for treating infections or bleeding of the genital-urinary tract, as well as dysentery. The fresh herb may also be applied topically to relieve sores and insect or snake bites on the skin. It is a common weed on campus.

马齿苋科马齿苋属
Portulaca oleracea (common purslane)
Portulacaceae

藜

藜科藜属
Chenopodium album
(white goosefoot)
Chenopodiaceae

又称灰菜。一年生草本，我国广泛分布。茎直立，具绿色或紫红色条棱，多分支。单叶互生，菱状卵形，嫩叶通常有紫红色粉末，叶缘具不整齐锯齿。花果期 5-10 月，花两性，簇生于枝条上，形成圆锥花序。果实为胞果，完全包被于花被内。校园荒地极多见，为著名杂草，幼苗叶可食。

Annual herb, distributed world-wide. This is a fast-growing weedy species, and cultivated in India as a food crop. It can also be used as animal feed. It is a common weed on campus.

铁苋菜

大戟科铁苋菜属
Acalypha australis
(southern copperleaf)
Euphorbiaceae

一年生草本，分布几乎遍于全国，长江流域尤多。雌雄同株。茎直立，多分支。单叶互生，椭圆状披针形，基出 3 脉；花期 5-7 月，穗状花序腋生，有叶状肾形苞片 1-3；雄花生于花序上部，雌花 1-3 朵生于基部苞片内。果期 7-11 月，蒴果，钝三棱形，有毛。杂草，校园路边及林边多见。

Annual herb, native to Asia. This is a weedy species, and a traditional herbal medicine for reducing inflammation and stopping bleeding. It is a common weed on campus.

粉花绣线菊

蔷薇科绣线菊属
Spiraea japonica (Japanese spiraea)
Rosaceae

落叶灌木，原产于日本，我国广泛栽培。株高1.5米，枝干光滑，或幼时具细毛，叶卵形至卵状长椭圆形，先端尖，叶缘有缺刻状重锯齿，叶背灰蓝色，脉上常有短柔毛；花期6-7月，簇生呈复伞花序，花淡粉红色至深粉红色，稀白色，雄蕊较花瓣长；花期7-9月，蓇葖果。未名湖南侧近有栽培。

Deciduous shrub, native to China, Japan and Korea with about eight varieties. This species is often planted as an ornamental landscape plant. It was recently cultivated south of Wei Ming Lake.

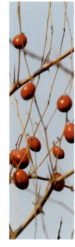

枣

鼠李科枣属
Ziziphus jujuba
(jujube or Chinese date)
Rhamnaceae

落叶乔木，我国特产，全国各地均有栽培。小枝成"之"字形弯曲。单叶互生，有锯齿，具3条主脉；托叶有时成刺。花期5-6月，聚伞花序腋生；花小，两性，黄绿色。果期9月。校园有该种的两个变种：枣（原变种 *Ziziphus jujuba* var. *jujuba*）和酸枣（*Ziziphus jujuba* var. *spinosa*）。枣的核果长圆形，深红色，味甜，枣核两头尖，为我国主要果树；校园不多，燕南园见有少数栽培。酸枣核果较小，圆形，味酸，枣核两头钝，酸枣富含维生素，可做饮料。该变种为暖温带阳生灌丛的标志性植物之一，校园多见。

Deciduous tree. This species is endemic to China. It is widely cultivated in other parts of the world with numerous cultivars. The flowers are an important nectar source for honeybees, the fruit is edible, rich in vitamin C, and often preserved or candied. The fresh and dried fruit are also used medicinally. Three varieties are recognized in *Flora of China*, and two of them are found on campus: *Ziziphus jujuba* var. *jujuba*, and var. *spinosa*. The variety *spinosa* has small round fruit with thin and sour mesocarp, and is very common on campus. The variety *jujuba* has bigger fruit with thick and sweet mesocarp, and is cultivated in Yan Nan Yuan.

多么不可思议，红润、饱满、甘甜、多产的枣，频频让人想起乡村里孩子的脸，想起天地的滋养，而竟然是由棘（酸枣）驯化而成的。这种多刺的灌木，在华北地区常常和荆条混生在一起。后来，人们就给它另起了名字，把"棘"中的两个"朿"从左右排列改成上下排列，这就是"棗"（"枣"的繁体）了，视觉上也是从棘刺并生变成硕果累累的样子。（刘凤）

石榴

石榴科石榴属
Punica granatum (pomegranate tree)
Punicaceae

落叶小乔木，原产于西亚南部。小枝平滑，一般有刺。单叶对生或簇生，长圆状披针形，全缘，两面光滑无毛。花期 5–6 月，花红色（栽培品种可为白色或黄色），萼片硬，肉质，花瓣倒卵形，5–8 枚或重瓣；雄蕊多数。果期 9–10 月，浆果，近球形，萼宿存；种子多数，具肉质多汁的外种皮（食用部分）和坚硬内种皮。著名果树和观赏植物，电教周边及燕南园有栽培。

Deciduous shrub or small tree. This species was probably introduced to China from Central Asia or Europe during the Han dynasty. It is widely cultivated in China for its showy flowers and delicious fruit (pomegranate), and is naturalized in West China. The pomegranate in Chinese culture has been considered an emblem of fertility and numerous progeny for the many-seeded fruit is a sign of fecundity. It is cultivated around the Audio-Visual Education Center and the Museum of Peking University History.

石榴本名"安石榴",据信是从中亚古国"安息"传入,传说是张骞通西域的时候引入中国的,而真正的引种者却茫然无考了。石榴因其奇异的果实被融汇到崇尚子孙繁盛的中国文化之中。(刘凤)

一叶萩

大戟科白饭树属
Flueggea suffruticosa
(shrubby flueggea)
Euphorbiaceae

名称来自日语。落叶灌木，我国分布广泛。雌雄同株或异株。茎丛生，多分枝，小枝绿色，纤细，有棱线。单叶互生，具短柄，叶片椭圆形，全缘，两面无毛。花期 5–7 月，花 3–12 朵簇生于叶腋；花小，无花瓣，萼片 5，淡黄色，卵形。果期 7–9 月，蒴果，3 浅裂。未名湖西部石桥北侧两边有 3 株。

Deciduous shrub, native to most parts of China, Japan, Korea and Far East Russia. This is a toxic plant, however the leaves and flowers of this species are used as medicine to treat infantile paralysis, neurasthemia, and facial paralysis. A few shrubs are found at the western end of Wei Ming Lake.

蒙椴

椴树科椴树属
Tilia mongolica (Mongolian linden)
Tiliaceae

落叶乔木，分布于我国北方。树皮红褐色，小枝光滑。单叶互生，卵圆形，基部心形或截形，常不对称，叶缘有锯齿。花期6-7月，聚伞花序，花序梗下部与1带状苞片合生；小花通常两性，花瓣5；雄蕊多数。果期9-11月，核果或浆果。为园林树种，俄文楼西侧林中有2-3株。

Deciduous tree, native to North China. The flowers are an excellent nectar source. Several trees are cultivated on campus.

牵牛

旋花科番薯属
Ipomoea nil (ivy morning glory)
Convolvulaceae

一年生缠绕草本，原产于美洲，归化中国，各地常见栽培。植株被毛。单叶互生，近圆形，常 3 裂至中部。花期 6–9 月，花序腋生；花两性，萼片披针形，不向外反曲，花冠蓝色或淡紫色，漏斗状。果期 7–10 月，蒴果，近球形，3 瓣裂。观赏植物，常逸为杂草，校园荒地路边有生长。

Annual twining herb. This species is native to South America, and now nearly circumtropically distributed. It is naturalized in China, and sometimes cultivated for its blue or light purple flowers. The seeds are used medicinally. It is a species of weed on campus.

打碗花

旋花科打碗花属
Calystegia hederacea
(Japanese false bindweed)
Convolvulaceae

一年生缠绕草本或有匍匐分支，我国各地有分布。植株无毛。茎上部叶三角状戟形，常2裂，叶基微心形，全缘。花期7-9月，花两性，单生叶腋，苞片大，2枚，包于萼片之外；花冠漏斗状，粉红色；雄蕊5。果期8-10月，蒴果，卵形，萼片宿存。常见杂草，校园有生长，镜春园内荒草地多见，水沟水塘边也有。

Annual or perennial herb, native to Asia. This species is very common in fields, wastelands, roadsides, river banks, and sandy soils. Its roots have been used to cure irregularity of menstruation, and its flowers could be used to ease a toothache. It is often seen on campus.

那个美丽的禁忌，我们小时候都听过。再次开始注意它是去年夏天，在物理楼墙边的小花坛，一大片大叶黄杨之间夹杂着几朵粉色小花，着实可爱。这么多年远在千里之外，想起大人当年的语调似乎还有些神秘。（张慧婷）

圆叶牵牛、牵牛、田旋花、打碗花的区别

圆叶牵牛和牵牛的花的柱头头状，3裂，田旋花、打碗花的花的柱头短条形，2裂。

圆叶牵牛的叶圆形，全缘，牵牛的叶3裂。圆叶牵牛的萼片卵形，牵牛的萼片卵状披针形，先端尾状。

田旋花的花梗上有2小形苞片，不靠近花萼。打碗花有2较大卵形苞片，紧贴着花萼基部着生。田旋花的叶常3裂，两侧裂片小，耳状，中裂片狭长。打碗花的叶片多5裂，中裂片卵状长圆形。

圆叶牵牛

旋花科番薯属
Ipomoea purpurea
(common morning glory)
Convolvulaceae

一年生缠绕草本，原产于南美，归化中国，南北各地分布广泛。茎被短柔毛和倒向的长硬毛。单叶互生，圆卵形，被糙伏毛，基部心形，边缘全缘。花期6-9月，花两性，花冠紫红色或粉红色（也有白色），漏斗状。果期9-10月，蒴果，近球形，3瓣裂；种子黑色。常见杂草，镜春园住宅篱笆上有生长。

Annual twining herb, native to Mexico and South America. This species is cultivated, and escaped and naturalized in most parts of China and other countries worldwide. The seeds are used medicinally. It is a common weedy species, and often seen on roadsides or as hedges in residential areas on campus.

田旋花

旋花科旋花属
Convolvulus arvensis (field bindweed)
Convolvulaceae

多年生草本，北京分布极为普遍。茎平卧或缠绕，有棱。叶片戟形，全缘或3裂，基部耳形。花期5-8月，花两性，常单生叶腋，花梗细弱，苞片小，2枚，线形，远离萼片；萼片5，被毛，花冠漏斗形，粉红色，5浅裂；雄蕊5。果期7-9月，蒴果，球形，无毛。常见杂草，常生于荒地，生物楼东侧楼南墙脚下有生长。

Climbing or creeping perennial herb. It is native to Eurasia, and is an unwelcomed weed for cultivated plants. The whole plant can be used medicinally. It is a common weed on campus.

栾树

无患子科栾树属
Koelreuteria paniculata
(panicled goldraintree)
Sapindaceae

落叶乔木，原产于我国中部和北部。1回或不完全2回奇数羽状复叶，互生，初生时为紫红色，通称木兰芽，可以食用，成叶的小叶长卵形，边缘具锯齿或裂片。花期6-7月，大型圆锥花序顶生；花小，金黄色。果期9-10月，蒴果三角状卵形，中空，顶端尖，成熟后红褐色或橘红色。花、果均美丽，为观赏树种，校园栽培多，且多老树，如北阁的西边、生物技术楼西北侧均有。

Deciduous tree or shrub, endemic to central and North China. It is widely cultivated elsewhere in the world as an ornamental because of the aesthetic appeal of its flowers, leaves and seed pods. It is widely cultivated on campus. This species is different from the Chinese golden rain tree in that its seed pods are green when they are young, and that it flowers in summer (June to July).

院中还有一棵大栾树……这树开小黄花，春夏之交，有一个大大的黄色的头顶，吸引了不少野蜂。……夏天的树，挂满浅绿色的小灯笼，是花变的。以后就变黄了，坠落了。满院子除了落叶还有小灯笼，扫不胜扫。专司打扫院子的老头曾形容说，这树真霸道。后来他下世了，几个接班人也跟着去了，后继无人，只好由它霸道去。看来人是熬不过树的。（宗璞）

荆条

马鞭草科牡荆属
Vitex negundo var. *heterophylla*
(five-leaved chastetree)
Verbenaceae

落叶灌木，北京分布普遍。小枝四棱。掌状复叶对生，具长柄，小叶 5-7，椭圆状卵形，边缘具切裂状锯齿，背面灰白色，被柔毛。花期 6-8 月，圆锥花序，花冠蓝紫色或偶为白色，二唇形，雄蕊 4，2 长 2 短，和花柱都稍外伸。果期 7-10 月，核果，球形或倒卵形。为华北地区常见的阳坡灌丛植物，也是良好的蜜源植物，未名湖周边的小山坡上和镜春园有生长。

Deciduous shrub or small tree, distributed in Asia, East Africa, and the Pacific Islands. Its flower is an important nectar source. The leaves and the essential oil of the plant are used to treat superficial bruises, injuries, and skin infections, and to reduce inflammation and swelling of joints due to rheumatism. It is distributed around Wei Ming Lake.

枸杞

茄科枸杞属
Lycium chinense (wolfberry)
Solanaceae

落叶灌木，我国各地有分布。枝条细弱，有纵条纹，小枝顶端锐尖成棘刺状。单叶互生或 2-4 枚簇生，卵形，基部楔形。花期 5-9 月，花两性，花萼通常 3 中裂或 4-5 齿裂，花冠漏斗状，淡紫色，5 深裂；雄蕊 5。果期 6-10 月，浆果，红色，卵状，可以食用。重要药用植物，校园各小山上有生长，尤以未名湖附近为多。

Deciduous shrub, native to Eurasia, on slopes, wastelands, saline places, roadsides, and near houses. This is a famous medicinal plant in China. It is widely cultivated in China, especially in the Ningxia Autonomous Region. Its fruit is used as a tonic, the root bark is used for relieving cough and reducing fever, and the young leaves are eaten as a vegetable. Dried wolfberries are often added to rice congee and almond jelly, as well as used in Chinese tonic soups. The berries are also boiled as an herbal tea, often along with chrysanthemum flowers and/or red jujubes. Some wines contain wolfberries. It is commonly seen on slopes around Wei Ming Lake.

纵有良药之名，却无大牌之招摇。春暖之时，枸杞隐藏在燕园的芸芸草木之中；秋收之日，未名湖边来往不绝的人却少有能发现它鼎鼎大名的果实的。这种以滋补阳气而闻名的朱红的果实却是如此的低调。（江都）

厚萼凌霄

紫葳科凌霄属
Campsis radicans (trumpet vine or trumpet creeper)
Bignoniaceae

落叶木质藤本，原产于美国，我国各地有栽培。羽状复叶对生，小叶 7-9，卵形，基部不对称，两面无毛，边缘有锯齿。花期 6-9 月，圆锥花序顶生，花橙红色，花萼钟形，5 裂至 1/3 处，花冠喇叭状。蒴果长如豆荚，顶端钝。种子多数。老生物楼西北边有一株。

There are only two species in this genus, one in China (Chinese trumpet vine) and one in the United States. This species is native to Southeast United States, and notable for its showy trumpet-shaped flowers. It is cultivated as a popular garden plant in China. A large plant is found on the northwestern side of the former Biology Building.

栝楼

葫芦科栝楼属
Trichosanthes kirilowii
(Mongolian snakegourd)
Cucurbitaceae

多年生草质攀缘藤本，原产于我国。雌雄异株。块根肥大，圆柱形。茎多分支，卷须细长。单叶互生，掌状3-7裂，边缘有疏锯齿，基部心形。花期7-8月，雄花成总状花序，雌花单生于叶腋，均为白色；花冠裂片5，边缘具流苏状细裂片。果期9-10月，瓠果，近球形，成熟时金黄色。药用植物，其根为"天花粉"，果为"栝楼"。药膳原址（网球场南侧）、博雅塔下草地或西侧山坡有生长。

Perennial climbing vine, native to China. This species has been used as a famous herbal medicine for hundreds of years in China. Its dried root "tian-hua-fen" contains an active compound trichosanthin - an inhibitor to certain viruses, but it could cause severe immune reactions. Its fruit is named as "gua-lou" and used as an expectorant. The female plants are found on southeast hills of Wei Ming Lake.

栝楼的果实有几样妙处,其一是成熟后,与茎牢牢相连,数年后也不会掉下来。其二是成熟后,果瓤籽粒黏在果实底部。任凭用手拨动果实,它总是迅速复位,如同不倒翁。其三是可做面膜。《古今图书集成》中说千年前的燕地女子冬月用栝楼涂面,至春暖方涤去,可使面白如玉。(刘华杰)

华北珍珠梅

蔷薇科珍珠梅属
Sorbaria kirilowii (Kirilow falsespiraea)
Rosaceae

落叶灌木，原产于我国华北。枝开展。奇数羽状复叶互生，小叶片对生，无柄或近无柄，披针形至卵状披针形，边缘有尖锐重锯齿，羽状脉明显。花期7–9月，顶生大型圆锥花序；花小，花瓣5，白色。果期9–10月，蓇葖果，长圆形。观赏植物，未名湖南岸山坡草地、外文楼北侧等地多栽植，在植株下常见有种子萌发的小植株。

Deciduous shrub, endemic to North China. This is a robust and fast-growing species. The white floral buds in big inflorescence look like pearls, hence its Chinese name. It is widely cultivated on campus around buildings.

莲

莲科莲属
Nelumbo nucifera (lotus)
Nelumbonaceae

又称荷花。多年生水生草本，原产于中国、印度、伊朗和大洋洲，现世界各地广为栽培。根茎通称藕，肥厚，横走地下，可以食用。单叶基生，叶片盾状着生，初生叶为浮水叶，后生叶则挺出水面。花期7-8月，花大，两性，粉红色或白色，花瓣多数；雄蕊多数。果期8-9月，坚果通称莲子，可以食用，嵌于膨松的花托（通称莲蓬）中。著名观赏植物，西校门南侧有面积较大的荷花池两处；季羡林先生门前（朗润园）清塘中原来栽的荷花，最早是季老带自湖北洪湖，后为周一良先生称为"季荷"。

Aquatic perennial, native to China, India, and Australia. This species is widely cultivated as an ornamental for its beautiful flowers, and as food for its seeds and underground rhizomes. In Buddhist symbolism, the lotus represents purity of the body, speech, and mind as if floating above the muddy waters of attachment and desire. In the classical written and oral literature of many Asian cultures, the lotus represents elegance, beauty, perfection, purity and grace. It is also used symbolically in poems and songs. The lotus ponds on campus are one of the attractions in summer.

鸣鹤园原有大片荷花,红白相间,清香远播。动乱多年后,寻不到了。现在勺园附近、朗润园桥边都有红荷……红荷的红不同于桃、杏,鲜艳中显出端庄,就像白玉兰于素静中显出华贵一样。我曾不解为什么佛的宝座做莲花状,再一思忖,无论从外貌或品德比较,没有比莲花更适合的了。(宗璞)

东方香蒲

香蒲科香蒲属
Typha orientalis (oriental cattail)
Typhaceae

多年生沼生高大草本,分布于我国长江以北。雌雄同株。地上茎粗壮。单叶基生,条形,长40—70厘米,宽0.4—0.9厘米;叶鞘抱茎。花果期5-8月,雌雄花序紧密连接,雄花序在上,雌花序在下。果序褐色,棒状;果实为小坚果,椭圆形至长椭圆形。生物技术楼西北有生长。

Perennial herb, native to China, some Asian countries, and Australia. This species usually grows by slowly moving waters. It is often planted on the edge of lakes or ponds as an ornamental. Its leaves are a resource of fiber, and its pollen is used as a traditional medicine. It is found in the lake west of the Biotechnology Building.

当海淀的稻田和沼泽已不复存在,水被人们禁锢在一方方水泥池里,北大自由流淌的水以及水边自由生长的植物是值得诉说的。它的果在秋天成熟,果序常被湖边游玩的顽皮少年像烛一般点燃,青烟带着袅袅的香。(吴岚)

木槿

锦葵科木槿属
Hibiscus syriacus (shrubalthea)
Malvaceae

落叶灌木或小乔木，原产于我国中部。茎直立，单叶互生，菱状卵形，3裂，具3主脉，边缘有粗锯齿。花果期7-9月，花两性，单生，具线形副萼，花瓣5或为重瓣，通常为红紫各色。蒴果长圆形，具毛。为常见观赏植物，花可食用，图书馆和办公楼周边都有种植。

Deciduous shrub. This species is native to China, and was cultivated very early and distributed to the Middle East along early trade routes. It is now cultivated in most tropical and temperate regions with many cultivars. The flowers are edible as a vegetable. The double flower cultivars are often seen on campus.

木槿花的三种颜色……最愿见到的是紫色的，好和早春的二月兰、初夏的藤萝相呼应，让紫色的幻想充满在小园中，让风吹走悲伤，让梦留着。（宗璞）

爬山虎

葡萄科地锦属
Parthenocissus tricuspidata
(Japanese creeper or grape ivy)
Vitaceae

又称地锦。落叶木质藤本，产于我国各地。卷须与叶对生，顶端分支，分支末端膨大为吸盘，植株即借此攀爬。叶互生，通常3裂或为三出复叶，秋季转为红色。花期6月，聚伞花序与叶对生；花小，多为两性，雄蕊与花瓣对生。果期9–10月，浆果，小球形，蓝黑色，被白粉。著名垂直绿化植物，校园栽培较多，如静园各院建筑墙上多有攀爬，秋季红叶甚美丽。

Deciduous woody vine, native to China, Japan and Korea. The leaves turn deeply red in the fall. It is widely grown as a climbing ornamental plant to cover the facades of masonry buildings. The roots and stems are used medicinally. The buildings around Jing Yuan are covered by this species.

北大的黛瓦青砖营造的是冷静的调子，冷碧的爬山虎会让很多建筑显得有些阴森。但一院到六院人字墙上的爬山虎却因开阔的静园，独享了朝朝暮暮的阳光。沧桑的十二面山墙上，生长着这些每年都有青春的植物。就如同十二张宣纸上，爬山虎如墨色，或横或斜，或皴或染，有时碧绿如泼，有时疏影婉约，是北大造景中的神来之笔。（王立刚）

臭椿

苦木科臭椿属
Ailanthus altissima
(tree of heaven ailanthus)
Simaroubaceae

古名樗。落叶乔木，分布于我国北方。雌雄同株或异株。树皮灰黑色，平滑，稍有浅裂纹。奇数羽状复叶大，互生；小叶13-41，卵状披针形，近基部具少数粗齿，齿顶有腺点，有臭味，中上部全缘。花期6-7月，圆锥花序顶生，花小，绿色。果期9-10月，翅果，长椭圆形，翅扁平膜质。校园很多，生物楼东北部路边山丘上有大树；且多见种子落地萌发而成的幼苗。

Deciduous tree, native to central and Northeast China. This species is a fast-growing tree, and introduced to Europe and North America in the 18th century. It was recorded in Li Shizhen's *Compendium of Materia Medica* in the Ming Dynasty, and almost every part of this tree is used in traditional Chinese medicine. Sometimes it is considered an invasive species because it suppresses competition with allelopathic chemicals in disturbed areas, colonizes quickly beyond its native range. It is a common tree species on campus.

臭椿那圆滚滚的枝条总带一点儿弯,显得皮实憨厚又有几分狡黠。庄子汪洋恣肆的词句里,樗木"大而无用",逍遥于远方无何有之乡。在略显局促的园子里,它也乐得长在无人打搅的墙根旁、土坡上,开一串琐碎又不显眼的花,依然自在且惬意。(蔡乐)

香椿

棟科香椿属
Toona sinensis (Chinese mahogany or Chinese toon)
Meliaceae

多年生落叶乔木，分布我国大部分地区。偶数羽状复叶互生，幼叶紫红色，有特殊香气，可以食用，成年叶绿色，叶背红棕色，叶柄红色，小叶 5–11 对，长椭圆形。花期 6 月，圆锥花序顶生，下垂，花小，两性，白色，有香味。果期 10–11 月，蒴果，红褐色，果皮革质。农园食堂北门、治贝子园、四院西南处有栽培。

Deciduous tree, native to China, and Southeast Asia. This is generally an upland species but also occurs at lower altitudes in China. It is planted for shade, and as an ornamental. The young shoots and leaves are edible as a delicacy in China. The trees are widely used medicinally. Several trees are planted in Zhi Beizi Yuan, and near the Nong Yuan Resturant.

香椿的树皮有纵裂纹，这点古人早知之：传说古代一皇帝到了农家，农家用香椿嫩叶炒鸡蛋招待皇帝，皇帝吃了大加赞赏，要封香椿为"树王"，他将"树王"牌子错挂在了臭椿树上，香椿树见了，气得树皮都纵裂了。故事为虚构却有趣！（汪劲武）

香椿似有一种南国的风情,羽状复叶总是优雅地舒展,尖端的小叶却娇羞般垂着,仿佛被看不见的雨水打湿了。若趁四月出芽时采摘,便是"叶香可啖"的佳肴,有"春来第一鲜"的美誉。俗话说,"房前一株椿,春菜常不断",这大概就是在农园食堂旁种植它的原因吧!(蔡乐)

香椿与臭椿

二者的区别最明显的是香椿的果为蒴果,臭椿的果为翅果。香椿的花为两性花,臭椿的花为杂性花(单性、两性花兼有)。

香椿的叶为偶数羽状复叶(偶尔出现奇数羽状复叶)小叶全缘(较幼的树小叶边缘有锯齿),臭椿的叶为奇数羽状复叶(偶尔出现偶数羽状复叶),小叶基部常有1浅裂片,裂片背面有一圆盘形腺点。这一特征十分稳定,是从叶上区分二者的可靠特征。

无花无果又无叶时,从树皮上区分明显,香椿主干树皮纵裂(条裂),臭椿树皮不条裂,且十分坚固。

梓

紫葳科梓属
Catalpa ovata (yellow catalpa or Chinese catalpa)
Bignoniaceae

落叶乔木，分布于我国长江以北。小枝具毛。单叶对生，有时3叶轮生，阔卵形，长宽近相等，基部心形，全缘或浅波状，或3-5浅裂，基部具掌状脉5-7条。花期6-7月，圆锥花序顶生；花两性，花冠黄白色，二唇形，内面具2条黄色条纹及紫色斑点；能育雄蕊2，退化雄蕊3。果期7-9月，蒴果，长圆柱形。陈守仁国际研究中心前有数株树，皆50年代栽植。

Deciduous tree. The species in genus *Catalpa* have a disjunction distribution pattern between East Asia and North American. This species is endemic to central, North, and Northeast China. It is widely cultivated as an ornamental. Its wood can be used to make a traditional Chinese instrument, and the whole plant is used medicinally. Several trees are found near the buildings of campus Radio Station, and the Tan Siu Lin Center for International Studies.

梓在今天的北方栽培得不多了，但是在古代却非常重要。梓木用处广泛，所以"梓人"即指木工。梓木还可以用做雕版印刷的木板，故古代把著作即将印行叫做"付梓"。《诗经》有"维桑与梓，必恭敬止"。意思是父母植于住宅周围的桑树和梓树，一定要必恭必敬，从此"桑梓""梓里"就成了家园、故乡的代称。梓材亦用作棺木，古人敬死，因为那才是所有人真正的故乡。（刘凤）

黄金树、楸树、梓树的区别

黄金树、楸树、梓树均属紫葳科梓属（*Catalpa*），该属物种呈东亚–北美间断分布，中国包括引进的物种共有 7 个，而在北京所能见到的三个物种均在燕园找到，其中还不乏百年以上的老树，实属难得。这三个物种的叶子和花部特征较为特异，容易区分：梓树叶子具 3–5 浅裂，黄金树和楸树均为全缘叶；楸树叶双面均无毛，而黄金树叶背面密布柔毛；楸树花有密集的紫色斑点，5 月份就开始开花，黄金树和梓树花期较晚，黄金树花为白色，梓树花为黄白色，且花序中花小而多。

黄金树

紫葳科梓属
Catalpa speciosa
(northern catalpa or cigar tree)
Bignoniaceae

落叶高大乔木，原产于美国中部。小枝无毛。单叶对生，大形，宽卵形，基部心形至截形，全缘，背面密生柔毛。花期6–8月，圆锥花序顶生；花两性，花冠白色，二唇形，内面具2条黄色及不甚明显的紫褐色条纹或斑点。果期7–9月，蒴果，长圆柱形。26楼南墙边和21楼南墙边有栽培。

Deciduous tree, native to Midwest United States. It is widely grown as an ornamental tree outside its native range, and also in China. The wood is an excellent material for carving, boatbuilding, and furniture. It is cultivated in front of Buildings No. 21 and No. 26.

挺括的叶片间是雪白的花荫，玲珑的冠筒里似镌有繁复的符箓，和它的名字一样，每一朵都是对未来美好生活的祈愿，仿佛在呼应远处博雅塔上的檐铃。（蔡乐）

忍冬

忍冬科忍冬属
Lonicera japonica
(Japanese honeysuckle)
Caprifoliaceae

又称金银花。落叶攀缘灌木，我国大部分地区有分布。幼枝密被黄褐色腺毛和短柔毛。叶纸质，长椭圆形，全缘，两面被毛。花期6-8月，花两性，成对生于叶腋；花冠二唇形，外面具柔毛，上唇具4裂片，下唇反转，初开时白色，后变黄色，极芳香；雄蕊5。果期8-10月，浆果，球形，黑色。著名观赏植物和药用植物，未名湖西侧小桥边有一株，镜春园内有栽种。

Deciduous twining shrub, native to East Asia. This species has become naturalized in America, Australia, as well as a number of Pacific and Caribbean islands, and is classified as a noxious weed in some countries. Its flowers have anti-bacterial and anti-inflammatory properties. One plant is found by the bridge at the western end of Wei Ming Lake, and it is also cultivated in Jing Chun Yuan.

忍冬原产东亚，因为花朵美丽芳香，在庭院广为栽培。然而，这样的一种香花，在美国人眼里却是大毒草。忍冬在引入美洲之后，成了一种令人头疼的入侵植物。或许这是另外一个版本的淮北为枳吧。（刘夙）

锦带花、忍冬、金银忍冬、白果毛核木的区别

锦带花的叶缘有浅锯齿,其他四种的叶均为全缘。锦带花花冠紫玫瑰色,内面灰白色,其他三种花冠均不为紫红色。锦带花的花1-4朵成伞形花序,生于短侧枝之顶,其他三种多为二花成对生叶腋。锦带花的果实为蒴果,白果毛核木的果实为核果,其他两种均为浆果。

忍冬为木质藤本,花冠长1-4厘米,金银忍冬为灌木,花冠长约2厘米。白果毛核木也为灌木,但其花冠长度不到1厘米。

蜀葵

锦葵科蜀葵属
Alcea rosea (common hollyhock)
Malvaceae

多年生草本，原产于我国。茎直立挺拔，丛生，不分支，全体被星状毛和刚毛。单叶大形，互生，叶片近圆心形，粗糙，两面均被星状毛。花期7–8月，长总状花序；花大而美丽，有红、黄、白等各种颜色，并有单瓣（花瓣5）和重瓣品种。常见观赏花卉，物理楼前栽培有各种花色的植株。

Perennial herb. This species originated in Southwest China and has been grown in Europe since at least the 15th century. It is a very special species since it is not known from any truly wild situations. It is introduced throughout temperate regions now with many cultivars, and it is cultivated as an ornamental and is used medicinally. Many plants with different colored flowers are cultivated outside the Physics Building.

梧桐

梧桐科梧桐属
Firmiana simplex
(Chinese parasol tree)
Sterculiaceae

落叶乔木，原产于我国南方以及日本。树皮青绿色，平滑。单叶互生，掌状3–5裂，基部心形，基生脉7条，叶柄与叶片等长。花期6–7月，圆锥花序顶生；花两性，淡黄绿色，萼片花瓣状。果期10月，蓇葖果5，叶状；种子球形。树形美丽，为很好的庭园观赏植物，但不耐风寒，以"梧桐落叶，一叶知秋"而著称，北方栽培较少。26楼门口有2株，老生物楼前、俄文楼西南草坪各有1株。

Deciduous tree, native to China and Japan. This species is widely cultivated in China and North America as an ornamental or shade tree. The wood is used for the soundboards of several Chinese instruments, and after processing the seeds are edible. In Chinese culture, it is a symbol of nobility and elegance and the only tree on which the legendary phoenix bird would land. It is also a symbol of loneliness, which has been cited in some famous ancient Chinese poems. Several trees are found in front of the Building No. 26, former Biology Building, and near the statue of Li Dazhao.

梧桐畏寒，一般只分布在南方，但燕园的梧桐树却长得枝繁叶茂，让来自南方的学子倍感亲切。尽管因气候的原因，这儿的梧桐树很难长出"一株青玉立，千叶绿云委"的气势，但它毕竟是传说中凤凰栖息的地方，难道真是应了此间的少年，藏着人中的龙凤？（顾红雅）

紫薇

千屈菜科紫薇属
Lagerstroemia indica (crape myrtle)
Lythraceae

落叶小乔木，原产于我国。树皮易脱落，树干光滑，用手抚摸，全株微微颤动，故俗名"痒痒树"。单叶互生或对生，近无柄，椭圆形。花期6-9月，长达百余日，故又名"百日红"，圆锥花序顶生；花大而美丽，花瓣6，粉红色、红色（红薇）或白色（银薇），边缘有不规则皱状缺刻，基部有长爪；雄蕊多数。果期7-9月，蒴果，椭圆状球形，6瓣裂。著名观赏植物，图书馆东草地西缘、勺园北边草地边缘有数株。

Deciduous shrub or small tree, native to China, and other countries in East, Southeast, and South Asia. This species is widely cultivated throughout the warm regions of the world as an ornamental in gardens and parks. It is also a good material for bonsai. Several plants are cultivated near the university Library, and north of Shao Yuan. It is also sporadically seen around campus.

紫薇的花期很长，有了它，盛夏的校园不再单调。它很奇特，不知经历过什么磨难，而没了"皮"，又是"千曲百折"的花瓣。紫薇又名痒痒树，据说挠它的树干，它就会觉得痒痒而有反应，校园里紫薇光滑的树干，都是被好奇的人们摸光的么？（顾红雅、吴岚）

燕园草木

古名莠。一年生草本，分布于我国南北各省。秆直立。叶舌毛状；叶片阔条形，无毛。花果期 6—10 月，圆锥花序圆柱状，通常绿色或褐黄色，形似狗尾；不育小枝退化为刚毛状，粗糙。果实为颖果。该物种是我国北方主要杂粮谷子 (*Setaria italica*) 的野生种。为常见杂草，校园多见，草地路边均有。

Annual herb. This species is native to Eurasia, but known on most continents as an introduced weed. It is the wild antecedent of the crop, foxtail millet, but a weed to many types of crops. It is very common on campus.

狗尾草

禾本科狗尾草属
Setaria viridis (green foxtail)
Poaceae

一万年前黄河流域的先民幸运地将狗尾草驯化成了今天我们还常常食用的小米（粟）。而别的一些杂草至今仍然只是杂草。遥想想万年前那些棕黄土地上黄皮肤的先民高高捧起一抔金黄的粟米，也算是我们瞥见那草间瓦上飘摇不定的卑微植物时应有的敬意。（刘夙）

多年生草本，全国均有分布。全株被毛。单叶互生，长椭圆形，基部稍狭，有时有小耳，半抱茎，全缘或具细锯齿。花期7-10月，头状花序顶生；总苞半圆形，总苞片数层，外层披针形，内层线状披针形或线形，干膜质；舌状花1层，黄色，管状花亦为黄色。果期8-11月，瘦果，长椭圆形。全草入中药。校园草地多野生。

Perennial herb, native to East Asia. The whole plant is used medicinally. It is a common weed on campus.

旋覆花

菊科旋覆花属
Inula japonica (Japanese inula)
Asteraceae

泽芹

伞形科泽芹属
Sium suave (water parsnip)
Apiaceae

多年生挺水草本，分布于我国长江以北。茎具明显纵棱，节间中空。1回奇数羽状复叶互生，小叶3-9对，小叶片条状披针形；浸入水中的植株茎下部生出2回羽状全裂的沉水叶，裂片细条形，锐尖。花果期7-9月，复伞形花序；花小，两性，花瓣5，白色。双悬果近球形。镜春园内水域多有生长。

Perennial semi-aquatic herb, distributed in North and East China, Japan, Korea, Russia and North America. This species has reputed medicinal value. It is commonly seen in the lakes in Jing Chun Yuan.

一串红

唇形科鼠尾草属
Salvia splendens (scarlet sage)
Lamiaceae

多年生草本，原产于巴西，野生种高度可达数米，其栽培品种多为矮化植株，常作一、二年生栽培，我国各地广泛栽培。茎直立，四棱形，光滑。单叶对生，卵形，边缘有锯齿。花期7—10月，轮伞花序具2—6花，密集成顶生假总状花序；花萼钟形，颜色多样，二唇形，上唇全缘，下唇2裂；花冠颜色多样，二唇形，下唇3裂，冠筒伸出萼外；雄蕊和花柱伸出花冠外。果期8—10月，小坚果，椭圆形，有3棱，平滑。校园花坛常见栽培。

Perennial herbs, native to Brazil. This species could grow to 1.5 to 8 meters tall in its native form. It has been hybridized and bred into many dwarf cultivars with different floral colors. The cultivar with bright red inflorescences is commonly planted on campus.

马蓼

蓼科蓼属
Polygonum lapathifolium
(willow weed)
Polygonaceae

又称酸模叶蓼。一年生湿生草本，我国分布广泛。茎直立，节部膨大。叶互生，披针形或宽披针形，全缘，具缘毛，上面常具马蹄形黑斑；托叶鞘膜质，无毛，先端截平。花期 6–7 月，圆锥花序；花两性，粉红色或淡绿色。果期 7–9 月，瘦果，扁卵圆形，包在宿存的花被内。镜春园水边成片分布。

Annual herb. This is a widely distributed weedy species, with several varieties recognized. It develops into large populations on the edge of the waters in Jing Chun Yuan.

红蓼

蓼科蓼属
Polygonum orientale
(prince's feather)
Polygonaceae

一年生湿生草本，我国分布广泛。植株高大，茎直立，节膨大，上部分支多。单叶互生，宽椭圆形，全缘，两面被毛；托叶鞘膜质，顶端向外反卷，具缘毛。花期7-9月，圆锥花序顶生或腋生，常弯曲下垂；花两性，红色或粉红色。果期9-10月，瘦果，近球形，包在宿存的花被内。京郊农村常栽培作观赏植物。镜春园水边多见。

Annual herb, distributed in Eurasia and Australia on roadsides, near houses, and wastelands. This species is usually cultivated in gardens. Stems and leaves can be used medicinally for relieving rheumatic disorders. It is commonly seen on watersides in Jing Chun Yuan.

乌蔹莓

葡萄科乌蔹莓属
Cayratia japonica (bush killer)
Vitaceae

多年生草质藤本，分布我国南部各省，北京有栽培。茎圆柱形，扭曲，有纵棱，多分枝，带紫红色；卷须二歧分岔，与叶对生。鸟足状复叶互生，小叶5，椭圆形，边缘具疏锯齿，中间小叶较大，侧生小叶较小。花期6-7月，花两性，4数，雄蕊与花瓣对生。果期7-8月，浆果，卵圆形。观赏植物，电教东南角有栽培。

Herbaceous vine, native to Asia and Australia. This species is usually distributed in disturbed areas, and sometimes cultivated as an ornamental. The whole plant is used as a herbal medicine to relieve swelling, and to enhance diuresis and detoxification. It is found at the southeastern corner of the Audio-Visual Education Center.

槐

豆科槐属
Sophora japonica (pagoda tree)
Fabaceae

又称国槐。落叶乔木，原产于我国，南北各省有栽培。树皮暗灰色。奇数羽状复叶互生，小叶 7–15，卵状长圆形。花期 7–8 月，圆锥花序顶生，有柔毛；花两性，蝶形，黄白色，略具芳香。果期 10 月，荚果，肉质，念珠状，不开裂。良好的园林树和行道树，但常有槐尺蛾为害。为北京市市树之一（另一种是侧柏）。校园栽培很多，南校门内马路两侧皆是。

Deciduous tree, native to East Asia. It is cultivated throughout China, also a popular ornamental tree in Europe, North America and South Africa. Many intergrading varieties and horticultural forms exist. Its flower is not only the nectar source for honeybees, but also used as traditional Chinese medicine. The pagoda tree is the municipal tree of Beijing. Many trees are planted on campus, with the most famous ones along the main entrance to the South Gate. Several big trees over 100 years old are found in Yan Nan Yuan.

主路两边的国槐，排列有序，树冠穹合，如同一条绿色的长廊，以前每年报到的新生都从这同一条路走进来，但毕业的时候却从各自的路散出去。（王立刚）

一年生湿生草本，全国各地均有分布。茎圆柱形，肉质。单叶互生，带肉质，卵状披针形，全缘，基部成膜质鞘。花果期6–10月，总状花序，佛焰苞有柄，心状卵形，边缘对合折叠。花两性，蓝色，花瓣3，常2大1小；发育雄蕊3。蒴果椭圆形，压扁状。朗润园湖边偶见。

Annual herb, native to China, and other countries in East and Southeast Asia. This species usually grows near water. It is used medicinally for febrifugal, anti-inflammatory, and diuretic effects; and for treating sore throats and tonsillitis. It is seen near the lakes in Lang Run Yuan.

鸭跖草

鸭跖草科鸭跖草属
Commelina communis
(Asiatic dayflower)
Commelinaceae

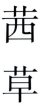

多年生攀缘草本，我国大部分地区有分布。茎四棱形，茎棱、叶柄、叶缘和叶中脉均有倒钩刺。单叶，4片（偶为8片）轮生，卵形至卵状披针形，叶脉5，弧形。花果期6-9月，聚伞花序；花小，花冠黄白色，5裂；雄蕊5。果实肉质，成熟时红色。其根部可提取红色物质，为人类最早使用的红色染料之一。常见杂草，校园多有野生。

Perennial twining herb. This weedy species is morphologically variable, and widely distributed worldwide. It was extensively cultivated from antiquity until the mid-nineteenth century for a red pigment derived from roots in China, many other regions of Asia, Europe and Africa. The root has also long been used medicinally. It is a common weed on campus.

茜草科茜草属
Rubia cordifolia (common madder)
Rubiaceae

燕园草木

在今天看上去平凡无奇的茜草，在染料缺乏的古代却是十分珍贵的天然资源。因其根部含有茜素和紫茜素，可以用来把织物染成红色或橙色。直到1868年，德国的两位化学家格莱伯（C. Gräbe）和利伯曼（C. Liebermann）人工合成了茜素之后，茜草种植产业就很快消失了。但这曾经艳丽过千百年裙裾的小草并不会寂寞。（刘夙）

求米草

禾本科求米草属
Oplismenus undulatifolius
(undulate-leaf oplismenus)
Poaceae

一年生草本，广泛分布于我国南北各省。秆较细弱，基部横卧地面，节处生根。叶鞘遍布刺毛，或仅边缘具纤毛，叶舌膜质；叶片披针形，与一般禾草有异，通常皱而不平，有横脉，通常有细毛或疣毛。花果期 7–11 月，圆锥花序紧缩。果实为颖果，椭圆形。校园路边常见，未名湖南土丘上更连片生长。

Perennial herb, widely distributed in warm-temperate and subtropical regions of the Northern Hemisphere, uplands of India and Africa. This species is characterized by its wrinkled leaves, which can be used as animal feed. It is usually found in large populations under trees on campus.

野大豆

豆科大豆属
Glycine soja (wild soybean)
Fabaceae

一年生草本。茎缠绕、细弱，疏生黄褐色长硬毛。羽状复叶互生，具3小叶，小叶卵圆形，全缘，基部近圆形，两面被毛。花期6–8月，总状花序腋生；花两性，蝶形，淡紫红色。果期7–9月，荚果，狭长圆形或镰刀形，两侧稍扁，密被黄色长硬毛，在种子间缢缩，通常含3粒种子。为栽培大豆的野生型，常见于镜春园水边。

Annual herb, native to China, Japan, Korea, and Russia. This species is protected in China because of its importance as the wild progenitor of soybean (*G. max*). The whole plant is used medicinally. It is sometimes seen in Jing Chun Yuan.

短尾铁线莲

多年生草质藤本，我国分布广泛。小枝紫褐色似铁丝。叶对生，1-2回羽状复叶或2回三出复叶，小叶5-15，长卵形，边缘疏生粗锯齿，偶3裂。花期7-8月，圆锥花序；花两性，直径1.5-2厘米，萼片4，开展，白色，无花瓣。果期8-10月，聚合瘦果，各具短的羽毛状宿存花柱。博雅塔附近土坡至临湖轩北山坡一带多有生长。

Perennial twining herb, native to central and North China, Korea, Mongolia, and Far East Russia. The stems of this species are used medicinally. It is very easily found around Wei Ming Lake when it is blooming.

毛茛科铁线莲属
Clematis brevicaudata
(short plume clematis)
Ranunculaceae

葎草

大麻科葎草属
Humulus scandens (Japanese hop)
Cannabaceae

一年或多年生缠绕草本，为全国广布的杂草。雌雄异株。茎枝和叶柄密生倒钩刺。单叶对生，掌状5-7深裂，两面具粗糙毛。花期7-8月，雄花组成圆锥状花序，花小，黄绿色；雌花序则近球形，腋生。果期9-10月，果序揉碎有略似啤酒花的气味。校园常见杂草。

Annual herb with twisting slender stems. It is native to East Asia, and naturalized in Europe and Eastern North America. This weedy species is difficult to control, partially because the female plants produce numerous seeds. The whole plant can be used medicinally and the seed oil to make soap. It is very common on campus.

益母草

唇形科益母草属
Leonurus japonicus
(Chinese motherwort)
Lamiaceae

古名蓷、茺蔚。二年生直立草本，分布我国各地。茎四棱形，被短柔毛。单叶对生，中部叶全裂，裂片全缘，上部叶渐小，形状也渐简单。花期 7–9 月，轮伞花序腋生；花冠粉红色，二唇形，上唇长圆形，下唇 3 裂；雄蕊 4。果期 9–10 月，小坚果 4。常见杂草，校园凡草地或者荒地皆有生长。

Annual or biennial herb, native to China, Japan, Korea, and Cambodia. It has escaped cultivation and become naturalized in other parts of the world. The whole plant is used medicinally, mostly for treating woman's diseases. Its Chinese name can be literally translated as "beneficial herb for mothers". It is commonly found on campus.

益母草似乎总是和女性联系在一起。《诗经》中有"中谷有蓷，暵其干矣。有女仳离，慨其叹矣。""蓷"即是益母草。诗人用山谷中的益母草被太阳晒干来起兴，深切地同情一位被丈夫抛弃的女子，感叹她的择偶不慎。和这种东方药草相映成趣的是，欧洲也有一种益母草，亦被视为能治妇女疾病的草药。（刘夙）

玉簪

百合科玉簪属
Hosta plantaginea (hosta)
Liliaceae

多年生宿根草本，原产于我国，北京栽培多。单叶基生成丛，卵形至心状卵形，基部心形，叶脉呈弧状。花期 6–8 月，总状花序顶生，高于叶丛；花白色，管状漏斗形，具浓香，在夜间开放，花被片 6。果期 8–10 月，蒴果，圆柱状，具 3 棱。观赏植物，一教北面路边有栽培。

Perennial herb, endemic to China. This species is widely cultivated as an ornamental with numerous cultivars. It is also used as a ground cover under shade. The whole plant is used medicinally. A big population is found behind the First Teaching Building.

玉簪花从不要人照料，只管自己蓬勃生长。……在晨光熹微或暮色朦胧中，一柄柄白花擎起，隐约如绿波上的白帆，不知驶向何方。有些植物的繁茂枝叶中会藏着一些小活物，会吓人一跳。玉簪花下却总是干净的。可能因为气味的缘故，不容虫豸近身。（宗璞）

萝藦

萝藦科萝藦属
Metaplexis japonica
(Japanese metaplexis)
Asclepiadaceae

古名芄兰。多年生草质藤本，广泛分布于我国南北各省。全株有白色乳汁。单叶对生，卵状心形，背面粉绿色，无毛；叶柄顶端有丛生腺体。花期7–8月，总状聚伞花序腋生；花白色，具副花冠，芳香。果期9–10月，蓇葖果大，单生。校园分布较多，常见攀爬于树木、栏杆之上。

Perennial twining herbaceous vine, widely distributed in China, Japan, Korea, and adjacent Russia. The stems and roots are toxic, and used as medicine for traumatic injury, snake bites, and infantile malnutrition due to intestinal parasites. It is a common weedy species on campus.

"萝藦"听起来似乎有点异域色彩，但其实在《诗经》中就有"芄兰之支，童子佩觿；虽则佩觿，能不我知？"芄兰即萝藦。觿是古代成年人的佩饰，形似萝藦的果实，所以诗人用芄兰来起兴。这首诗古人通常认为它讽刺了幼年即位的卫惠公对大臣"骄而无礼"。但也许它其实是一位年轻女子嗔怪恋人装模作样的抱怨之辞吧。（刘夙）

紫茉莉

一年生草本，原产于热带美洲，我国各省均有栽培。茎直立，光滑，分支多。单叶对生，卵形或卵状三角形，全缘，两面无毛。花果期7—10月，花两性，单生于枝的顶端，有多种颜色（可为红色、粉红色、黄色、白色或杂色），漏斗状。瘦果球形，黑色，具棱，形似地雷。校园各处见有栽培，为观赏花卉。

Annual herb, native to tropical America. This species was originally introduced as an ornamental, now naturalized in some areas of China. It has many cultivars with different floral colors. The whole plant is also used medicinally. It is cultivated near residential areas on campus.

紫茉莉科紫茉莉属
Mirabilis jalapa (four o'clock flower)
Nyctaginaceae

虎尾草

禾本科虎尾草属
Chloris virgata (feather fingergrass)
Poaceae

一年生草本，我国南北各省均有分布。秆丛生，稍扁，基部膝曲。叶鞘光滑无毛，上部叶鞘肿胀而包裹花序；叶片条形，扁平。花期 7–11 月，穗状花序簇生茎顶，呈指状排列；小穗紧密排列于穗轴一侧。果期 11–12 月，果实为颖果。典型的夏雨型短命植物，校园荒地多见。

Annual herb, native to many of the warmer temperate, subtropical, and tropical regions of the world. This is a widespread and very variable, weedy species on farm lands, but an excellent animal feed. It is common on roadsides on campus.

芦苇

禾本科芦苇属
Phragmites australis (common reed)
Poaceae

多年生水生或湿生高大草本，广泛分布于我国。茎直立，节下常生白粉。叶舌有毛；叶片长条形或长披针形，排列成两行。花果期7–11月，圆锥花序大形，分枝疏散，稍下垂，下部枝腋具白柔毛。果实为颖果，长圆形。未名湖边、镜春园水边均有，生物技术楼西湖边亦有一片。

This is a cosmopolitan, robust perennial. It grows in moist places along river banks and lake margins, forming large colonies. Its leaves are used as fiber and animal feed, and the whole plant could be used medicinally. The dense white fruiting bodies are very beautiful under gentle wind or moon light, inspiring many ancient poets in China. It is commonly seen on the edges of the lakes on campus.

它是《诗经》里在水一方的苍苍蒹葭，是建安诗文里满身秋霜的劲草，是达摩祖师渡江时足底的一叶。而在燕园里数个大大小小的湖边都长有芦苇，甚至有池鹭、白鹭、鸳鸯前来藏身。这是种至平淡而至神秘的植物。（张慧婷）

盒子草

葫芦科盒子草属
Actinostemma tenerum
(lobed actinostemma)
Cucurbitaceae

一年生草本。雌雄同株。枝纤细，卷须分为2岔，与叶对生。单叶互生，卵状心形，边缘有锯齿，基部心形，两面几无毛。花期7–9月，雄花序总状，腋生，雌花单生；花萼和花冠裂片均为5，披针形至线状披针形。果期9–11月，蒴果，绿色，从近中部盖裂。朗润园水域（干涸湖西北部）有野生。

Annual herb with slender branches. This species grows on watersides in China, India, Japan, Korea, and Southeast Asia. The whole plant and seeds are used medicinally as a diuretic and an antiedemic. It sometimes grows into a large population along watersides on campus.

燕园北部的十三公寓,季羡林先生晚年住在这里,坐北朝南,门口正对着朗润湖。2010年湖里缺水,湖中的植物反而更茂盛,当我沿曲折的小径继续向西时,觉得湖中青草上开小白花的草质藤本植物有些异样。细看之后,惊喜地发现它是葫芦科藤本植物盒子草。盒子草,也称合子草,中文名朴素而精准。每粒果实中只含两枚种子,所谓"合子",想必《本草纲目拾遗》称之"鸳鸯木鳖",也是基于此。关于名字,还有另一种解释。此植物果实中间有一道缝合线,将果实分为上下两部分,好似一个小盒儿装着两枚宝贝。生命进化出所有如此精致的结构,着实令人叹服。(刘华杰)

多花胡枝子

豆科胡枝子属
Lespedeza floribunda
(many-flower bush clover)
Fabaceae

小灌木，分布于我国大部地区。枝条细长柔弱，先端下垂。三出复叶互生，小叶片小而薄，倒卵形或狭长倒卵形，全缘，下面密被白色绒毛。花期7–8月，总状花序腋生；花两性，蝶形，粉红色至紫色。果期8–9月，荚果椭圆形，有毛。在北京低海拔地区普遍分布，校园也有野生生长，鸣鹤园及镜春园附近小山坡常见。

Small subshrub, native to China, India, Pakistan; and naturalized in Japan. This species has a very strong root system and a symbiotic relationship with nitrogen-fixation bacteria. It is sometimes used as ornamental plants in gardens. It is often seen on the hills near Ming He Yuan and Jing Chun Yuan.

白果毛核木

忍冬科毛核木属
Symphoricarpos albus
(Common snowberry)
Caprifoliaceae

也叫白雪果。落叶灌木，原产于北美。单叶对生，椭圆形，羽状脉，全缘，背面有毛。花期7-8月，花小，花冠粉红色，钟形。果期9-11月，果实球形，亮白色。观赏植物。信息科学技术学院门口有引种栽培。

Deciduous shrub. This species is native to North America. It is often planted as a cover for game in its native range. The brightly white berries make the plant a good candidate for ornamental purposes. It was recently introduced to campus.

一年生草本，世界广布，北京分布极为普遍。单叶互生，卵形，全缘或有不规则波状齿。花期7-9月，4-10朵花组成短蝎尾状花序；花冠白色，5裂；雄蕊5。果期8-10月，浆果，球形，黑色，具宿存的花萼，可以食用。校园野生很多，路边草地林下皆可见。

Annual herb, widely distributed in Asia and Europe, and introduced to America, Australasia and South Africa. The plant has been used for food and medicine in many countries. Its fruit is edible, and its leaves could be used to treat ulcers in mouth and to inhibit the growth of certain tumors. It is a common weed on campus.

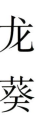

龙葵

茄科茄属
Solanum nigrum
(European black nightshade)
Solanaceae

凤尾兰

龙舌兰科丝兰属
Yucca gloriosa (Spanish dagger)
Agavaceae

常绿木本，原产于北美，我国广泛栽培。叶螺旋排列于茎端，幼龄植物的叶因茎短而似基生；质坚硬，有白粉，剑形，顶端硬尖，边缘光滑，老叶边缘有时具疏丝。花期6-10月，圆锥花序，花大而下垂，花被片6，乳白色，常带红晕。果期8-10月，果实下垂，椭圆状卵形，不开裂。观赏植物，一体西北部路边、老生物楼东侧草地、英杰交流中心东侧均有栽培。

Evergreen shrub, native to the coast and barrier islands of southeastern North America. Its showy white flowers and robust evergreen leaves make it a popular ornamental for gardens worldwide. It was recently cultivated on the eastern parts of campus.

多年生草本,分布我国南北各省。茎直立,密被短柔毛。单叶互生,下部叶有长柄,2回羽状深裂,中部叶羽状深裂,裂片披针形,上部叶全缘,所有叶下面均密被灰白色短毛。花期7-8月,头状花序多数。果期8-9月,瘦果,无冠毛,长不及1毫米。校园常见野草。

Perennial herb, native to North and Northeast China, Japan, Korea, and Far East Russia. This is a weedy species, and the whole plant is used medicinally. It is a common weed on campus.

野艾蒿

菊科蒿属
Artemisia lavandulifolia
(lavender leaf wormwood)
Asteraceae

牛膝

苋科牛膝属
Achyranthes bidentata
(ox knee achyranthes)
Amaranthaceae

又称怀牛膝。多年生草本，我国除东北外均有分布。茎四棱形，节膨大。单叶对生，椭圆形，两面有毛，先端具尾尖。花期7–9月，穗状花序腋生或顶生；花两性，在后期反折；苞片宽卵形，小苞片刺状。果期9–10月，胞果，椭圆形。实验西馆门外两侧林下有生长，系70年代栽培后转为野生。根可入中药。

Perennial herb, native to Asia. This species is a Chinese herbal medicine. In Nepal its root juice is used to relieve toothache. Its seeds have been used as a substitute for cereal grains during famines. It escaped cultivation and has become a common weed on campus.

黄花蒿

菊科蒿属
Artemisia annua (annual wormwood)
Asteraceae

一年生草本，分布于全国各省。茎直立，具纵沟，无毛。单叶互生，有浓烈气味，中部叶轮廓卵形，2回至3回羽状深裂，上部叶渐小，常为1回羽状深裂。花果期8-10月，头状花序球形，总苞无毛，2-3层，内层宽膜质；花全为管状花，黄色。果实为瘦果，无冠毛。校园极多，荒地路边草地多见。

Annual herb, native to temperate Asia, but naturalized throughout the world. The plant has been used by Chinese herbalists in ancient times to treat fever, but it is most famous for curing malaria. In 1971, Chinese scientists demonstrated that the plant extracts had antimalarial properties, and in 1972 the active ingredient, artemisinin, was isolated and its chemical structure described. It is now commonly used in tropical nations where malaria is prevalent. It is a common weed on campus.

气味浓烈的黄花蒿在上世纪 70 年代给了中国药学一个惊喜,人们受东晋葛洪《肘后备急方》中"青蒿一握,以水二升渍,绞取汁,尽服之"可治疟疾的记载的启发,从黄花蒿(青蒿)中分离出了青蒿素,成为中国科学家原创的极少数几种现代药物之一。(刘凤)

一至二年生草本，分布于我国和其他东亚国家。全株疏生短毛，茎直立，上部多分枝。单叶互生，叶片卵形至长椭圆形，有粗钝锯齿。花果期5-6月（二年生时）或7-10月（一年生时），头状花序大形，单生枝顶；野生原种舌状花1-2轮，呈浅堇至蓝紫色，栽培品种花色丰富，但尚未育出浓黄色；管状花黄色。果实为瘦果，楔形，有易脱落的冠毛。观赏花卉，校园花坛年年有栽培。

Annual or biennial herb, native to China, Korea and Japan. This species was first introduced to France in the 18th century, and is now becomes an important ornamental flower in the world with numerous cultivars of various floral colors. It is cultivated on campus.

翠菊

菊科翠菊属
Callistephus chinensis (China aster)
Asteraceae

一年生小草本，原产于北美，现归化中国。全株被毛，茎直立，有分枝。单叶对生，全缘或近全缘，基出3脉。花果期7–10月，头状花序小，舌状花5个，白色，管状花黄色。果实为瘦果。杂草，校园多见，凡路边、草地皆有分布。

Annual herb, native to America. This species is naturalized in China and other parts of the world, and considered as noxious weeds in some countries. Its young leaves are edible, and the whole plant is used medicinally as an astringent or for reducing inflammation. It is a common weed on campus.

牛膝菊

菊科牛膝菊属
Galinsoga parviflora (gallant soldier)
Asteraceae

鳢肠

菊科鳢肠属
Eclipat prostrata (false daisy)
Asteraceae

一年生湿生草本，全国各省皆有分布。全株被毛，茎、叶折断后有墨水样汁液。单叶对生，椭圆状披针形，无柄或基部叶有柄。花果期6-9月，头状花序腋生或顶生，总苞片2轮，5-6枚，有毛，宿存；舌状花白色，全缘或2裂，管状花白色。舌状花瘦果四棱形，管状花瘦果三棱形，表面都有瘤状突起，无冠毛。西校门水池附近草地有分布。

It is distributed in moist places as a weed all over the world. A black dye obtained from its leaves and stems is used for dyeing hair and tattooing. It is usually found near the lakes on the western part of campus.

婆婆针

菊科鬼针草属
Bidens bipinnata (Spanish needle)
Asteraceae

又称鬼针草。一年生草本，全国皆有分布。单叶对生，2回羽状深裂，小裂片三角形，边缘具不规则的锯齿，两面被疏柔毛。花果期8–10月，头状花序，总苞杯状；舌状花1–3朵，黄色，管状花黄色。果实为瘦果，线形，顶端有2–4芒刺，具倒刺毛。常见杂草，校园荒地路边多见，尤以镜春园内为多。

Annual herb. This plant is considered a weed in some tropical habitats. However, in some parts of the world it is a source of food or medicine. It is very common along roadsides on campus.

狼杷草

菊科鬼针草属
Bidens tripartita (three-lobe beggarticks)
Asteraceae

一年生草本，我国南北各省分布。单叶对生，叶柄有狭翅，中部叶常3-5裂，裂片狭披针形。花果期8-10月，头状花序，球形或扁球形；总苞片2层，内层披针形，干膜质，外层披针形，比头状花序长，叶状；无舌状花，管状花黄色。果实为瘦果，两侧边缘各有一列倒刺毛，顶端具2芒刺。杂草，镜春园水边多见。

Annual herb, native to large parts of the Northern Hemisphere. This species is a worldwide weed. Its leaves and roots are used medicinally. It is often seen near the lakes in Jing Chun Yuan.

海州常山

马鞭草科大青属
Clerodendrum trichotomum
(harlequin glory bower)
Verbenaceae

落叶灌木或小乔木，原产于我国，北京公园常见栽培。单叶对生，宽卵形或卵形，全缘，有时边缘有锯齿，叶柄长。花果期6-11月，花两性，花萼蕾时绿白色，后变成紫红色，宿存，花冠白色或带粉红色；花丝和花柱同伸出花冠之外。果实为核果，成熟时蓝紫色。观赏植物，老生物楼前西侧、行政楼西侧各栽有两株。

Deciduous shrub or small tree. This species exists in China, India, Japan, and Korea. It is often cultivated for its showy flowers, red calyx and blue fruit when mature, and for medicinal purposes. It is found west of the Administration Building, and by the former Biology Building.

257

羽叶栾树

无患子科栾树属
Koelreuteria bipinnata (Chinese goldraintree or Chinese flame tree)
Sapindaceae

又称全缘叶栾树，落叶乔木，分布于我国中部和南部地区，现全国栽培多；小枝暗棕色，密生皮孔；2回羽状复叶，互生，小叶7-11，薄革质，长椭圆形，顶端渐尖，全缘或偶有锯齿，两面无毛或沿中脉有短柔毛；花期8-9月，花黄色；萼片5，边缘有小睫毛；花瓣5，雄蕊8，花丝有长柔毛；果期10-11月，蒴果椭圆形，幼时紫红色。东校门新生命科学楼西面路边栽有一排。

Deciduous tree, endemic to central and Southeast China. This species is one of the few trees blooming in summer. Its big yellow inflorescence and reddish young seed pods are quite attractive. It is cultivated world-wide as a shade and ornamental tree, and is also used medicinally for treatments of conjunctivitis and epiphora. A few trees are cultivated west of the new Life Science Building.

山马兰

菊科马兰属
Kalimeris lautureana
(mountain kalimeris)
Asteraceae

多年生草本，分布于我国北方。茎直立，具纵棱纹，被毛。单叶互生，较厚，全缘或有疏锯齿，两面具毛。花果期7-9月，头状花序排成伞房状；总苞半球形，总苞片3层，上部绿色，无毛；舌状花淡蓝色，管状花黄色。果实为瘦果，倒卵形，冠毛极短。校园少见，校史馆东北侧草地有一小居群。

Perennial herb, native to North China, Korea, and Far East Russia. The whole plant is used medicinally. Only a few populations are found along the roadsides on west campus.

甘菊

菊科菊属
Dendranthema lavandulifolium
(lavandulate-leaf dendranthema),
Asteraceae

多年生草本，广泛分布于我国。茎直立或斜生，被灰白色绵毛。单叶互生，叶片矩圆形或卵形，羽状深裂，裂片2-5对，每个裂片又2-5浅裂或深裂。花果期9-10月，头状花序单生于花梗上；舌状花和管状花均为黄色。果实为瘦果，无冠毛。校园极多，尤以未名湖南岸林下一带为多，秋季开花时为校园一景。

Perennial herb, native to China, India, Japan, Korea, and Mongolia. The extract of the flowers could inhibit certain viruses and some pathogenic bacteria. The flowers could be used as herbal medicine to lower blood pressure. It is very common on campus, and its yellow flowers accentuate pleasant scenery in the fall.

雪松

松科雪松属
Cedrus deodara (Himalayan cedar)
Pinaceae

常绿高大乔木，原产于喜马拉雅山地西部，现世界各地广泛栽培。雌雄同株。树干挺直，枝条平展稍下垂，形成秀美的塔形树冠。叶针形，横切面三角形。花期10–11月，雄球花大，长卵圆形，雌球花小，卵圆形。次年种子成熟，球果直立，但在北京很少结实。为世界五大庭园树种（南洋杉、雪松、金钱松、日本金松、巨杉）之一，校园有栽培。

Evergreen tree, native to extreme southwest of Xizang, and other countries west of the Himalayas. This species is widely cultivated for its pagoda-shaped branches and upright trunk. The timber is utilized in shipbuilding, furniture, bridges, and construction. Several trees are planted on campus, the most famous ones are on the lawn west of the Administration Building. After a snowfall, they are popular locations for taking photos.

雪松披着庄严肃穆的青衫,每一片叶子都洒上了一层闪亮的银粉,勾勒出深浅不一的绿色,恰似墨绿的海洋泛起晶莹的浪花。我不由自主地想起那首诗:未名湖是个海洋……据说雪松的原产地是喜马拉雅山麓,那层莹白是来自那道雪线吗?眼前的雪松已经长出银绿色的雄球花,再过不久,它周围的土地就会被风中飞扬的花粉染成金黄,染得恰似老家那清明前后滚上松花的金团糕点。(蔡乐)

索引

巴天酸模	*Rumex patientia*	82
白　杜	*Euonymus maackii*	125
白　杄	*Picea meyeri*	60
白果毛核木	*Symphoricarpos albus*	245
白皮松	*Pinus bungeana*	54
半　夏	*Pinellia ternata*	162
斑种草	*Bothriospermum chinense*	71
抱茎小苦荬	*Ixeridium sonchifolium*	73
暴马丁香	*Syringa reticulata*	155
萹　蓄	*Polygonum aviculare*	164
草地早熟禾	*Poa pratensis*	159
侧　柏	*Platycladus orientalis*	65
臭　草	*Melica scabrosa*	109
臭　椿	*Ailanthus altissima*	200
刺儿菜	*Cirsium setosum*	112
翠　菊	*Callistephus chinensis*	252
酢浆草	*Oxalis corniculata*	158
垂　柳	*Salix babylonica*	10
大花糯米条	*Abelia × grandiflora*	140
大花野豌豆	*Vicia bungei*	92
打碗花	*Calystegia hederacea*	177
德国鸢尾	*Iris germanica*	102
地　黄	*Rehmannia glutinosa*	76
棣棠花	*Kerria japonica*	86
点地梅	*Androsace umbellata*	49
东方香蒲	*Typha orientalis*	194
东京樱花	*Cerasus yedoensis*	22
杜　仲	*Eucommia ulmoides*	19
独行菜	*Lepidium apetalum*	103
短尾铁线莲	*Clematis brevicaudata*	230
多花胡枝子	*Lespedeza floribunda*	244
二球悬铃木	*Platanus acerifolia*	116
粉花绣线菊	*Spiraea japonica*	169
风花菜	*Rorippa globosa*	160
凤尾兰	*Yucca gloriosa*	247
附地菜	*Trigonotis peduncularis*	72
甘　菊	*Dendranthema lavandulifolium*	260
杠　柳	*Periploca sepium*	124
枸　杞	*Lycium chinense*	186
构　树	*Broussonetia papyrifera*	46
狗尾草	*Setaria viridis*	214
栝　楼	*Trichosanthes kirilowii*	188
海州常山	*Clerodendrum trichotomum*	257
旱　柳	*Salix matsudana*	12
盒子草	*Actinostemma tenerum*	242
红丁香	*Syringa villosa*	126
红花锦鸡儿	*Caragana rosea*	128
红　蓼	*Polygonum orientale*	220
红瑞木	*Cornus alba*	119
厚萼凌霄	*Campsis radicans*	187
胡　桃	*Juglans regia*	66
虎尾草	*Chloris virgata*	239
互叶醉鱼草	*Buddleja alternifolia*	123
华北珍珠梅	*Sorbaria kirilowii*	190

263

中文名	学名	页码	中文名	学名	页码
华山松	*Pinus armandii*	53	牛膝	*Achyranthes bidentata*	249
槐	*Sophora japonica*	222	牛膝菊	*Galinsoga parviflora*	253
黄菖蒲	*Iris pseudacorus*	101	女贞	*Ligustrum lucidum*	154
黄刺玫	*Rosa xanthina*	87	欧洲荚蒾	*Viburnum opulus* subsp. *opulus*	122
黄花蒿	*Artemisia annua*	250	爬山虎	*Parthenocissus tricuspidata*	198
黄金树	*Catalpa speciosa*	206	平车前	*Plantago depressa*	113
黄栌	*Cotinus coggygria*	50	婆婆针	*Bidens bipinnata*	255
黄杨	*Buxus sinica*	52	蒲公英	*Taraxacum mongolicum*	40
灰栒子	*Cotoneaster acutifolius*	115	七叶树	*Aesculus chinensis*	68
茴茴蒜	*Ranunculus chinensis*	165	牵牛	*Ipomoea nil*	176
火棘	*Pyracantha fortuneana*	121	茜草	*Rubia cordifolia*	225
锦带花	*Weigela florida*	129	楸	*Catalpa bungei*	84
金银忍冬	*Lonicera maackii*	130	求米草	*Oplismenus undulatifolius*	228
荆条	*Vitex negundo* var. *heterophylla*	184	忍冬	*Lonicera japonica*	208
君迁子	*Diospyros lotus*	131	日本晚樱	*Cerasus serrulata* var. *lannesiana*	24
蜡梅	*Chimonanthus praecox*	6	日本小檗	*Berberis thunbergii*	83
狼杷草	*Bidens tripartita*	256	乳浆大戟	*Euphorbia esula*	106
藜	*Chenopodium album*	167	三裂绣线菊	*Spiraea trilobata*	135
鳢肠	*Eclipat prostrata*	254	三色堇	*Viola tricolor*	18
连翘	*Forsythia suspensa*	4	桑	*Morus alba*	44
莲	*Nelumbo nucifera*	191	山马兰	*Kalimeris lautureana*	259
流苏树	*Chionanthus retusus*	138	山桃	*Amygdalus davidiana*	2
龙葵	*Solanum nigrum*	246	山皂荚	*Gleditsia japonica*	141
芦苇	*Phragmites australis*	240	山楂	*Crataegus pinnatifida*	136
栾树	*Koelreuteria paniculata*	182	芍药	*Paeonia lactiflora*	80
萝藦	*Metaplexis japonica*	236	蛇莓	*Duchesnea indica*	111
葎草	*Humulus scandens*	231	石榴	*Punica granatum*	172
马齿苋	*Portulaca oleracea*	166	柿	*Diospyros kaki*	132
马蓼	*Polygonum lapathifolium*	219	蜀葵	*Alcea rosea*	210
马蔺	*Portulaca oleracea*	100	栓皮栎	*Quercus variabilis*	70
毛白杨	*Populus tomentosa*	14	水杉	*Metasequoia glyptostroboides*	58
毛梾	*Cornus walteri*	118	太平花	*Philadelphus pekinensis*	147
毛泡桐	*Paulownia tomentosa*	134	桃	*Amygdalus persica*	26
毛洋槐	*Robinia hispida*	120	铁苋菜	*Acalypha australis*	168
玫瑰	*Portulaca oleracea*	166	田旋花	*Convolvulus arvensis*	181
美国红梣	*Fraxinus pennsylvanica*	69	卫矛	*Euonymus alatus*	146
蒙椴	*Tilia mongolica*	175	蝟实	*Kolkwitzia amabilis*	142
牡丹	*Paeonia suffruticosa*	78	乌蔹莓	*Cayratia japonica*	221
木槿	*Hibiscus syriacus*	196	乌头叶蛇葡萄	*Ampelopsis aconitifolia*	107
泥胡菜	*Hemistepta lyrata*	161	梧桐	*Fimiana simplex*	211
拟南芥	*Arabidopsis thaliana*	35	夏至草	*Lagopsis supina*	110

香茶藨子	*Ribes odoratum*	48
香 椿	*Toona sinensis*	202
小花扁担杆	*Grewia biloba* var. *parviflora*	150
杏	*Armeniaca vulgaris*	30
荇 菜	*Nymphoides peltata*	148
萱 草	*Hemerocallis fulva*	156
旋覆花	*Inula japonica*	216
雪 柳	*Fontanesia philliraeoides* subsp. *fortunei*	127
雪 松	*Cedrus deodara*	261
鸭跖草	*Commelina communis*	224
洋 槐	*Robinia pseudoacacia*	88
野艾蒿	*Artemisia lavandulifolia*	248
野大豆	*Glycine soja*	229
一串红	*Salvia splendens*	218
一叶萩	*Flueggea suffruticosa*	174
异穗薹草	*Carex heterostachya*	108
益母草	*Leonurus japonicus*	232
银 杏	*Ginkgo biloba*	56
迎春花	*Jasminum nudiflorum*	1
油 松	*Pinus tabuliformis*	62
榆 树	*Ulmus pumila*	8
榆叶梅	*Amygdalus triloba*	28
虞美人	*Papaver rhoeas*	96
羽叶栾树	*Koelreuteria bipinnata*	258
玉 兰	*Yulania denudata*	32
玉 簪	*Hosta plantaginea*	234
郁 李	*Cerasus japonica*	25
鸢 尾	*Iris tectorum*	98
元宝枫	*Acer truncatum*	42
圆 柏	*Juniperus chinensis*	64
圆叶牵牛	*Ipomoea purpurea*	180
圆叶鼠李	*Rhamnus globosa*	114
月季花	*Rosa chinensis*	151
杂种鹅掌楸	*Liriodendron chinense* × *L. tulipifera*	144
早开堇菜	*Viola prionantha*	16
早园竹	*Phyllostachys propinqua*	104
枣	*Ziziphus jujuba*	170
泽 芹	*Sium suave*	217
中华小苦荬	*Ixeridium chinense*	74
皱皮木瓜	*Chaenomeles speciosa*	38
诸葛菜	*Orychophragmus violaceus*	20
梓	*Catalpa ovata*	204
紫丁香	*Syringa oblata*	36
紫花地丁	*Viola philippica*	17
紫 荆	*Cercis chinensis*	90
紫茉莉	*Mirabilis jalapa*	238
紫 藤	*Wisteria sinensis*	93
紫 薇	*Lagerstroemia indica*	212
紫玉兰	*Yulania liliiflora*	34

后记

　　记得是多年以前的事了，当我在古韵飘逸的未名湖边看到花团锦绣，春色满园，便萌生了把校园里这些有灵性的花草树木记录下来的念头。一次偶然的机会，我跟北大前任校长、植物生理学家许智宏院士提及此事，没想到竟然与他的想法不谋而合，许校长也是早就想要出一本有关校园的花草树木的图册了，于是我们就开始行动起来。值得一提的是，对校园植物有研究的不仅有生科院，还有其他院系的老师，比如刘华杰老师是哲学系的教授，他对植物的认识具有专业的水平，他拍摄的植物照片不仅艺术感强，而且对植物各种特征的展示清晰到位；张莹老师来自国际合作部，她是北大数学系本科毕业的，她对自然之美有一种天然的爱好，她所拍摄的每一幅植物照片都很简洁、唯美；王立刚老师毕业于哲学系，现在北大出版社工作，是本书的责任编辑，他的那篇著名的散文《北大最美的十棵树》激发了大家的灵感；刘夙本科和硕士研究生阶段分别就读于北大化学与分子工程学院和历史系，现为中科院植物所博士研究生，他系统地拍摄了校园植物的照片，并曾有过出一本《北京大学植物志》的雄心壮志。生科院的老师更是积极参与，饶广远教授是"植物学"课程的主讲老师，对北大校园里的植物十分了解，感情很深；魏丽萍教授虽然是生物信息学专业的老师，但她看植物的视角很特别，所拍摄的照片布局构思精巧，充满了油画的质感；瞿礼嘉教授现在主要做植物分子生物学的研究，他在植物分类方面得到了汪劲武先生教诲，对校园里的几种艾蒿属植物研究颇有心得。吴岚是生科院的研究生，赵瑞白、张慧婷、江都、蔡乐是生科院的本科生，他们利用业余时间为本书收集了资料、撰写了小短文、提供了精美的照片。

　　顾问汪劲武先生是我们的学术领袖，校园的所有植物都装在他的头脑里，你只要告诉

他一个地点和简单的植物特征，他马上可以告诉你那是什么物种；更让人感动的是，他经常骑着一个老牌的永久自行车在校园中转，我们常常接到他自称为"通风报信"的电话：哪儿的树开花了、哪儿的果熟了、哪儿挂上了满树的红果、哪儿有一个新的外来物种……，这对我们的工作是一种无形的鞭策和督促。另一位顾问是植物所的李振宇老师，他利用春节假期，将我们所有的植物照片校对了一遍，保证了物种的准确性。我们的工作还得到了众多高手的帮助，如生科院教学中心的孟世勇老师为本书提供了很专业的物种描述；生科院的顾孝诚教授得知我们在编一本有关校园植物的书时，主动提出为我们修改英文稿，并亲切地将校园植物称为她的老朋友；*Flora of China*（英文版的中国植物志，http://flora.huh.harvard.edu/china/index.html）的中方主编、中科院植物所的洪德元院士耐心解答了我们提出的有关物种归类的问题。李岩松副校长和国际合作部的夏红卫部长对本书也提出了很多建设性的建议，周曼丽女士帮助我们联系外教修改英文稿。特别感谢生科院的贺新强老师以及陈佳佳、蒙皓和藏潇同学为本书补充了一些珍贵的照片。有了许校长的总体指挥、汪先生和李老师的学术把关、编委会成员平时的积累和积极参与，以及众多热爱校园植物老师们的指点，才使得从编委会的成立到将书稿交给出版社，只用了短短的9个月时间，这是一个多才、多艺、快乐、高效的集体。

　　本书的植物分类系统、物种描述和学名主要参考*Flora of China*，但我们保留了极个别植物的常用名，如*Arabidopsis thaliana*在*Flora of China*为"鼠耳芥"，但大家均称其为"拟南芥"，因此我们就采用了拟南芥。因考虑到本书有科普书籍的特点，书中介绍了物种所属的科和属，但省略了其拉丁学名中的定名人。为了便于阅读，书中物种基本上是按照其在校园里的开花时间排的序，并在描述中加入了其在校园中主要的分布地点。

　　燕园植物的美是要用心去体会的，我们在本书中加入了一些著名学者、校友对校园植物的描写，也添加了一些编委对部分植物的鉴赏和理解，尽量将校园文化融入到校园植物的描述中，希望这种尝试得到大家的认可。本书是我们献给北京大学113周年校庆的礼物，时间上有些赶，恳请读者指出本书的不足，甚至错误之处，以便我们在再版时加以改正。

顾红雅
2011年5月4日